The Stargazer's Handbook

The definitive field guide to the night sky

GILES SPARROW

Quercus

Contents

Origin of the Constellations

Our planet is surrounded by a Universe that stretches for billions of light years in every direction. Wherever we look, stars, gas and dust clouds and distant galaxies are scattered more or less at random across space. For thousands of years, constellations have played a key role in helping us make sense of this apparent chaos.

Today's system of 88 constellations has evolved since prehistoric times: the earliest depiction of Taurus, the Bull, comes from a 17,000-year-old painting at Lascaux, France. The first lists of constellations ancestral to our own were compiled in ancient Mesopotamia more than 4,000 years ago.

Probably the earliest constellations to be recognized were those of the zodiac – a group of a dozen star patterns traditionally attributed special significance because the Sun's annual path around the sky passes through them. The word 'zodiac' comes from a Greek term meaning 'circle of animals', and indeed all but one of these 12 constellations represents a living creature. The exception, Libra, was once part of neighbouring Scorpius. Thanks to the almost flat alignment of their orbits around the Sun, the planets are also usually found within these constellations – the idea that the positions of these 'wandering' celestial bodies in the zodiac had some relationship to events on Earth dates back to at least the first millennium BC, and survives in the persistent fascination of astrology. Thanks to long-term cycles of change in Earth's axis of rotation, however, the Sun's present path across the sky is largely out of step with the astrological zodiac.

While different cultures around the world developed their own star patterns at different points in history, and many of these are still in local use, astronomers today use a standardized list developed from the work of the Greek-Egyptian astronomer and geographer Ptolemy of Alexandria. Around AD 150, Ptolemy summarized the astronomical knowledge of the classical world in a great work known as the *Almagest*. Ptolemy's list of 48 constellations includes all of the ancient zodiac signs, and most of the others still recognized in northern and equatorial skies – it survived unaltered for around 1,400 years.

It was only from the 16th century that European explorers returned the first details of the far southern stars.

Astronomers soon began to introduce these discoveries into their charts, often making further additions to northern skies at the same time. A dozen new constellations were added to the southern sky by the followers of Dutch navigators Pieter Dirkszoon Keyser and Frederick de Houtman around 1600. However, large areas of the sky remained forsaken until the French astronomer Nicolas Louis de Lacaille invented a further 14 constellations for his great survey of the southern stars, published in 1763.

Finally, between 1922 and 1930, the International Astronomical Union agreed a list of 88 constellations, each formally defined as an area of the sky rather than a group of stars. Together, these 88 encompass the entire celestial sphere, ensuring that every object in the sky now belongs to one constellation or another.

Within the constellations, astronomers have ordered stars in various ways. Traditionally, Greek letters are used to indicate the brightest stars, according to the system devised by German astronomer Johann Bayer in 1603. Fainter stars are often assigned 'Flamsteed numbers' (introduced by English astronomer John Flamsteed in the early 18th century), while unusual stars and other celestial objects (often collectively known as 'deep-sky' objects) bear numbers or letters from a variety of other systems.

The brightness of stars, meanwhile, is measured by their 'apparent magnitude' – the brighter a star appears, the lower its magnitude number. This system originated in ancient Greece, but was formalized in the 19th century: today a difference of one magnitude between stars indicates that one star is roughly 2.5 times brighter than the other. The brightest stars have negative magnitudes, and the faintest naked-eye stars are around magnitude 6.0. On the maps used in this book, stars of different sizes are used to represent different brightnesses – more information about interpreting the maps is supplied overleaf.

This book offers a comprehensive introduction to understanding the individual constellations, and the night sky. The introductory pages describe how our location on Earth affects our view of the Universe, and how astronomers impose an ordered system of coordinates and constellations onto the sky. The main section of the book then takes us on a journey through the 88 officially recognized star patterns, describing the wonders that lie within them that can be seen with the naked eye, binoculars or relatively modest amateur telescopes.

The constellations were the first tools of astronomy, and still offer an important means of ordering the cosmos. Even today, an ability to recognize these celestial patterns and a knowledge of their bright stars and significant objects is a vital tool in the arsenal of any amateur stargazer. Only learn to understand the constellations, and the entire book of the Universe will open up before you.

How to Use the Star Charts

On the maps used throughout the book, stars of different sizes are used to represent different brightnesses. The colour of stars on the maps represents their actual measured colours, usually indicative of their surface temperature.

The brightness of stars is measured according to a system known as magnitude, formalized in 1850 by the English astronomer Norman Robert Pogson. According to his scheme, the lower a star's magnitude number, the brighter it is. The key shown below represents the scales of stars used in the large individual maps.

On the individual maps, non-stellar 'deep-sky' objects are also represented using the special symbols shown in the lower key. While the stars shown on the maps should all be visible to the naked-eye under dark skies, most of the deep-sky objects will require either binoculars or a small telescope.

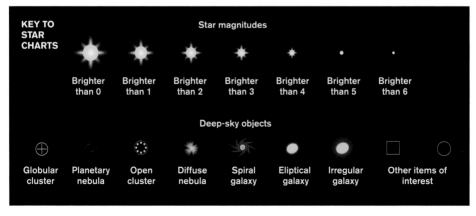

KEY TO STAR CHARTS

Star magnitudes

Brighter than 0	Brighter than 1	Brighter than 2	Brighter than 3	Brighter than 4	Brighter than 5	Brighter than 6

Deep-sky objects

Globular cluster	Planetary nebula	Open cluster	Diffuse nebula	Spiral galaxy	Eliptical galaxy	Irregular galaxy	Other items of interest

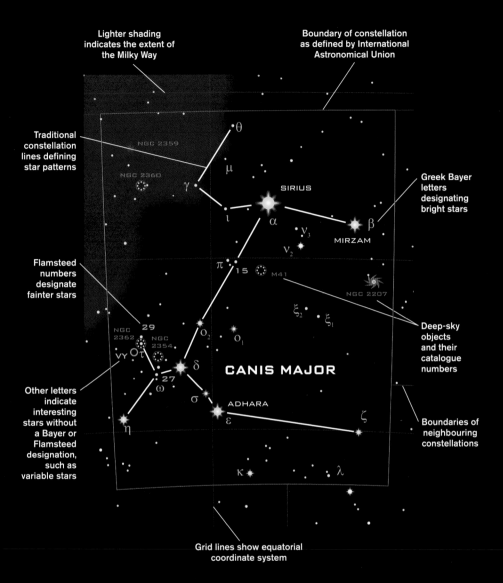

Lighter shading
indicates the extent of
the Milky Way

Boundary of constellation
as defined by International
Astronomical Union

Traditional
constellation
lines defining
star patterns

NGC 2359

NGC 2360

θ

μ

γ

SIRIUS

Greek Bayer
letters
designating
bright stars

ι

α

β

MIRZAM

ν₃

ν₂

Flamsteed
numbers
designate
fainter stars

π

15

M41

NGC 2207

ξ₂

ξ₁

Deep-sky
objects
and their
catalogue
numbers

NGC
2362

29

NGC
2354

ο₂

ο₁

VY

τ

δ

CANIS MAJOR

27

ω

Other letters
indicate
interesting
stars without
a Bayer or
Flamsteed
designation,
such as
variable stars

σ

ADHARA

ε

ζ

Boundaries of
neighbouring
constellations

η

κ

λ

Grid lines show equatorial
coordinate system

The Greek alphabet

The brightest stars in each constellation are represented
by a system of Greek 'Bayer letters', in accordance with the
Greek alphabet as shown here. However, the system has
many inconsistencies, with some stars bearing the 'wrong'
letters due to a variety of historical accidents. Fainter stars
down to the limits of naked-eye visibility are often given
'Flamsteed numbers', linked to their location within the
particular constellation.

α	alpha	ι	iota	ρ	rho
β	beta	κ	kappa	σ	sigma
γ	gamma	λ	lambda	τ	tau
δ	delta	μ	mu	υ	upsilon
ε	epsilon	ν	nu	φ	phi
ζ	zeta	ξ	xi	χ	chi
η	eta	ο	omicron	ψ	psi
θ	theta	π	pi	ω	omega

The Earth in Space

Ancient astronomers believed that Earth was the centre of the cosmos, a fixed location in space around which all the celestial objects – the Sun, Moon, planets and stars – moved. At first glance, this idea seems quite intuitive, but the reality of the Universe is rather different.

The first cosmological theory emerged from ancient Greece. It envisaged a static Earth surrounded by the Sun, Moon, planets and stars moving on a series of concentric spheres. Refined by the astronomer Ptolemy around AD 150, this 'geocentric' model went unchallenged for another millennium.

Then, in 1543, Nicolaus Copernicus marshalled the evidence for a 'heliocentric' Universe in which Earth was just one of several planets orbiting the Sun. Around 1610, this shift in our cosmic perspective was confirmed through the work of Johannes Kepler and the early telescopic observations of Galileo Galilei. Kepler realized that the problems with a heliocentric Universe could be resolved if the planets followed elliptical orbits rather than perfect circles.

A further diminution in Earth's stature arose from the realization that the stars were suns like our own, seen across immense distances. By the 1800s, larger telescopes allowed astronomers to begin mapping stars across our Milky Way galaxy. Today we understand that the Sun is just one insignificant star among several hundred billion, and that the Milky Way itself is one of countless similar systems across the Universe.

The immensity of cosmic distances, and the limitations of the speed with which even light can travel across space, mean astronomy is a largely observational science. Through analysis of starlight astronomers have discovered a surprising amount from our corner of space. Their studies have revealed the billion-year life cycles of stars, the spectacular conditions in which they are born and the strange remnants left behind when they die. They have shown the ways in which galaxies maintain their elegant structures, the processes by which solar systems and planets form, and have even offered clues to the origins and fate of the Universe itself.

But in order to get to grips with the nature of the cosmos, it is first necessary to understand the complex ways in which our location on Earth affects our view of the wider Universe.

The Visible Sky

From any point on Earth we can see roughly half of the sky – the rest is hidden by the bulk of the planet that lies beneath our feet. The precise region of the sky we see depends on both the planet's rotation and our particular location on Earth and in space (*see* page 12).

The three diagrams (below) show the very different skies seen by observers situated at Earth's north pole, at mid-latitudes and at the equator. As Earth spins on its axis, the segment of the sky visible for the mid-latitude and equatorial observers changes through the course of each day.

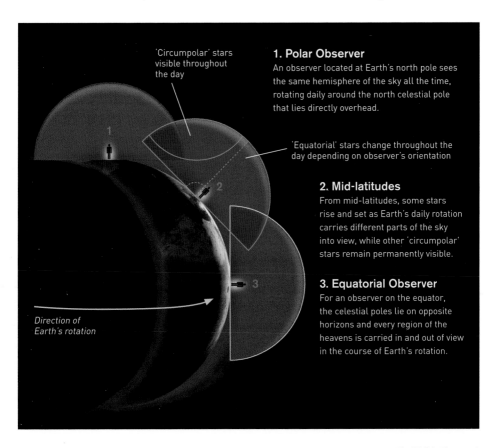

'Circumpolar' stars visible throughout the day

1. Polar Observer
An observer located at Earth's north pole sees the same hemisphere of the sky all the time, rotating daily around the north celestial pole that lies directly overhead.

'Equatorial' stars change throughout the day depending on observer's orientation

2. Mid-latitudes
From mid-latitudes, some stars rise and set as Earth's daily rotation carries different parts of the sky into view, while other 'circumpolar' stars remain permanently visible.

3. Equatorial Observer
For an observer on the equator, the celestial poles lie on opposite horizons and every region of the heavens is carried in and out of view in the course of Earth's rotation.

Direction of Earth's rotation

The Seasons

The tilt of Earth's axis relative to the plane of the Solar System produces seasons that affect the length of day and night and the areas of the sky that we can see throughout the year. It also creates much longer-term changes in the orientation of the sky and the direction of the celestial poles.

As Earth makes its annual journey around the Sun, its axis of rotation is tilted relative to 'upright' by some 23.5 degrees. The axis remains pointed in the same direction in space throughout the year and, as a result, the opposing hemispheres receive different amounts of sunlight from month to month (*see* diagram opposite).

In June, around the time of the northern 'summer solstice' when the days are longest, the northern hemisphere points towards the Sun and benefits from the Sun's longer, higher track across the sky. Meanwhile, in the southern hemisphere the days are short, the Sun's path through the sky is relatively low and its heating effect is reduced. Six months later, at northern winter solstice, the situation is reversed. Midway between these extremes, in March and September, both hemispheres receive equal amounts of sunlight at the vernal (spring) and autumnal equinoxes.

Midnight Sun
The effect of seasonal changes becomes more extreme at higher latitudes (regions around the equator, in Earth's tropical zones, receive more or less equal sunlight throughout the year). Above the Arctic and Antarctic Circles (lines of latitude at 66.5 degrees north and south), the Sun does not rise at all around the winter solstice, and never sets around the summer solstice – a phenomenon known as the midnight Sun. The north and south poles themselves experience the most extreme version of this, with six months of perpetual daylight in summer followed by an equal period of bitterly cold darkness in winter.

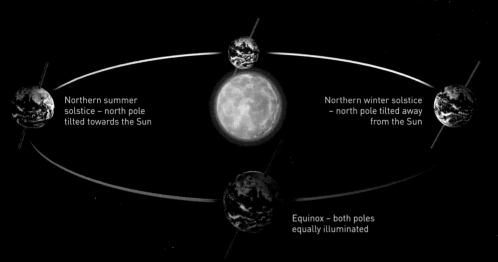

Northern summer solstice – north pole tilted towards the Sun

Northern winter solstice – north pole tilted away from the Sun

Equinox – both poles equally illuminated

Seasonal Change

As Earth orbits around the Sun each year, the amount of sunlight reaching different parts of Earth's surface varies as its axis of rotation maintains the same direction in space. The difference is at its most extreme around the solstices, while at the intermediate equinoxes, both hemispheres receive equal sunlight.

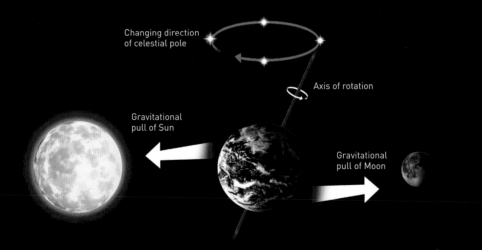

Changing direction of celestial pole

Axis of rotation

Gravitational pull of Sun

Gravitational pull of Moon

Precession

Although Earth's tilted axis remains pointed in the same direction from year to year, the gravitational forces of Sun and Moon tugging at the slight 'bulge' around our planet's equator cause its axis to wobble in a cycle that lasts some 25,800 years. As a result of this cycle, known as precession, the celestial poles describe a track around the sky and the Sun's position at the equinoxes tracks slowly westwards. One long-term consequence of this is that the direction of the celestial pole and the positions of stars in the widely used 'equatorial' coordinate system (see page 13) change slowly over time.

The Celestial Sphere

Although the true nature of the cosmos is very different, a simple model that imagines stars and other astronomical objects fixed on a transparent sphere around the Earth can be used as the basis for developing celestial measurement systems and mapping the skies.

The celestial sphere is an imaginary construct that flattens all the three-dimensional complexity of the Universe onto a sphere analogous to Earth itself. Instead of considering the reality of an Earth rotating on its axis once a day and spinning through space throughout the year (*see* page 11), this model reverts to the ancient idea of a static Earth, with a spherical shell of stars rotating around it daily, and the Sun and planets moving more slowly across this shell.

As illustrated on the opposite page, the celestial sphere pivots around points known as the celestial poles (extensions of Earth's north and south poles), and can be divided into hemispheres by a line known as the celestial equator. The Sun's annual path through the sky traces a line called the ecliptic, crossing the equator at two points corresponding to the vernal and autumnal equinoxes. The ecliptic marks the centre line of the region where the planets and Moon are typically found. It passes through the 12 traditional zodiac constellations.

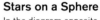

Stars on a Sphere
In the diagram opposite above, the near side of the celestial sphere has been removed for clarity. The celestial poles and celestial equator are directly aligned with Earth's own poles and equator, with latitude-like lines of declination running parallel to the celestial equator, and longitude-like lines of right ascension (R.A.) between the poles. The ecliptic (the Sun's path against the stars) is tilted at an angle to the celestial equator, crossing it in two places. The point where the Sun enters the sky's northern hemisphere at the northern spring equinox is known as the First Point of Aries (marked by the symbol ♈). It defines 0° declination, 0h right ascension.

Coordinate Systems
Just as for locations on Earth, measuring the positions of objects in the sky requires a fixed frame of reference. The simplest way of doing this, known as the alt-azimuth system, uses an observer's local horizon (*see* diagram opposite below). However, Earth's daily rotation means that alt-azimuth coordinates are unique to a specific location and time. In practice, 'equatorial' coordinates, which measure positions relative to fixed lines on the celestial sphere, are far more useful.

Celestial Sphere

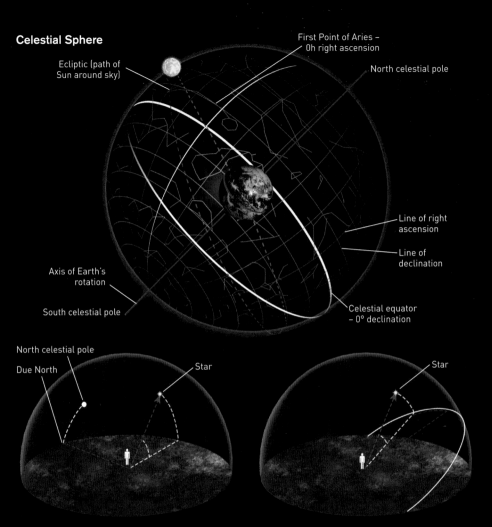

Ecliptic (path of Sun around sky)

First Point of Aries – 0h right ascension

North celestial pole

Line of right ascension

Line of declination

Axis of Earth's rotation

South celestial pole

Celestial equator – 0° declination

North celestial pole

Due North

Star

Star

Alt-azimuth Coordinates

The position of a star (or any other object) is measured according to its angle above the horizon (altitude) and its direction clockwise from due North (azimuth). An object's alt-azimuth location changes depending on the observer's location on Earth and the time of night they are observing.

1 Azimuth (angle clockwise from due North)

2 Altitude (angle above horizon)

Equatorial Coordinates

In this system, a star's position is described in terms of its declination (its angle in degrees north or south of the celestial equator) and its right ascension or R.A. (measured in hours, minutes and seconds east of 0h right ascension as defined by the First Point of Aries).

1 Declination (angle from celestial equator)

2 Right ascension (angle east of First Point of Aries)

3 First Point of Aries

4 Celestial equator

Defining Constellations

The constellations started off as simple patterns in the sky, made by joining up the brightest stars visible to the naked eye. But as new techniques revealed a wealth of new astronomical objects, that definition has had to change.

Constellations made simply by linking bright stars raised a few obvious issues. What if people disagreed about the shape of the pattern and the stars included? Did nearby stars outside of the pattern itself count as part of the constellation? And what about stars that were discovered later? As astronomical instruments improved, the situation was complicated by the need to catalogue more individual objects – unique names were all very well for the brightest stars, but impractical for every star in the sky. In 1603, Johann Bayer introduced a systematic scheme, listing stars with a combination of a Greek letter (normally corresponding to their relative brightness) and the possessive form of the constellation's Latin name.

So Sirius, the brightest star in Canis Major, becomes Alpha Canis Majoris. Later catalogues attempted to number stars by brightness or position, but every scheme has its failings, and in practice celestial objects are named in a variety of different ways.

A different point of view

The pattern of stars in a constellation is usually nothing more than a line-of-sight effect – the result of a chance alignment of objects, often at widely varying distances from Earth. For example, the five bright stars that form the distinctive 'W' of Cassiopeia actually lie (from east to west) 442, 99, 550, 229 and 54 light years from Earth respectively.

The brightness of a star seen from Earth, called its 'apparent magnitude', depends on both its true brightness or 'luminosity', and its distance (dwindling according to an 'inverse-square law' relationship. In other words, if two stars of the same apparent magnitude lie at 10 and 20 light years from Earth, the more distant must be four times as luminous.

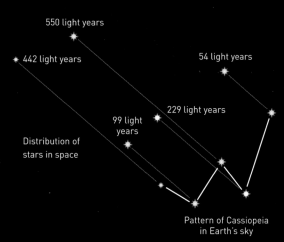

550 light years

442 light years

54 light years

Distribution of stars in space

99 light years

229 light years

Pattern of Cassiopeia in Earth's sky

A changing constellation

The stars of Taurus (1) are one of the most recognizable patterns in the sky, identified as the form of a bull by many cultures around the world (2). In the traditional 'star pattern' interpretation of constellations (3), bright stars are linked by lines and given Greek letter or number designations. In the official system of the International Astronomical Union, however, the constellation is a region of the celesial sphere demarcated by boundary lines and includes everything contained within it (4).

Mapping the Heavens

Broadly speaking, the celestial sphere can be divided into six large segments – two circumpolar regions and four zones around the celestial equator. Their visibility depends on an observer's location on Earth, and the time of night and year.

Circumpolar stars

The discs at the top of each page show the 'circumpolar' regions around the celestial poles. These are permanently visible for inhabitants in the middle latitudes of one hemisphere, and forever out of sight for their counterparts at similar latitudes in the other hemisphere.

Northern Hemisphere

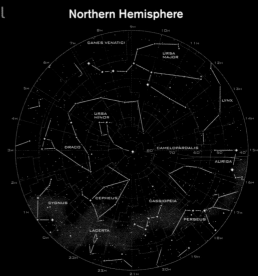

0h Right Ascension   **18h Right Ascension**

Southern Hemisphere

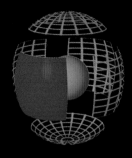

Seasonal skies

The segments along the bottom of the page show regions around the celestial equator whose visibility changes depending on the Sun's position on the ecliptic (*see* schematic above). Stars centred around 0h R.A. are best seen in early evening around October, those around 6h in January, those around 12h in April, and those around 18h in July.

12h Right Ascension

6h Right Ascension

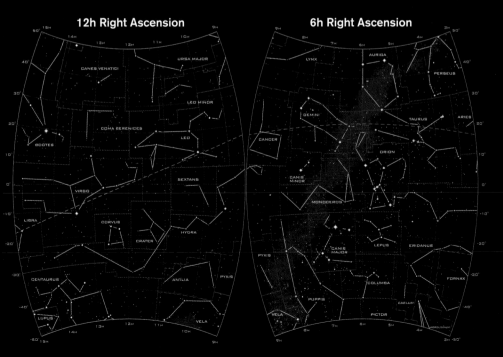

Ursa Minor

AT A GLANCE

NAME Ursa Minor
MEANING The little bear
ABBREVIATION UMi
GENITIVE Ursae Minoris
R.A. 15h 00m
DEC. +77° 42'
AREA 256 (56)
BRIGHTEST STAR Polaris (α)

The northernmost constellation in the sky, Ursa Minor is famed for its brightest star, Polaris. The pole star, which is found at the end of the Bear's 'tail', lies within half a degree of the north celestial pole and remains almost perfectly still in the sky even as the rest of the heavens spin around it.

Ursa Minor consists of seven moderately bright stars whose pattern strongly resembles the larger and brighter shape of the Plough, or Big Dipper, in the nearby Great Bear Ursa Major (*see* page 32). This resemblance is probably the reason why the constellation eventually became known as the Lesser Bear, though others have seen it as a hunting dog or even an extended wing of the mighty dragon Draco. Polaris, the constellation's brightest star with a magnitude of 2.0 and a distance of some 2,400 light years, is a rare example of a star that has changed noticeably in recorded history. It is an entire magnitude brighter than it was in ancient times and until recently was a pulsating variable star – its changes in size and brightness have only dwindled away over the past century.

AT A GLANCE BOXES

NOTE The item heading 'Area' in the **At A Glance** boxes contains two numbers for each constellation. The first figure gives the size of the constellation in square degrees, while the second figure (in brackets) is the size ranking from 1 (biggest) to 88 (smallest).

Polar trails
Long-exposure photographs of the region around the north celestial pole capture the sky's apparent rotation over a period of up to an hour. All the stars in the sky describe arcs of light as Earth spins beneath them, but the arcs are shortest for those stars closest to the pole, including the pole star Polaris.

Draco

AT A GLANCE

NAME Draco
MEANING The dragon
ABBREVIATION Dra
GENITIVE Draconis
R.A. 15h 09m
DEC. +67° 00'
AREA 1,083 (8)
BRIGHTEST STAR Eltanin (γ)

The long, sinuous figure of Draco, the celestial dragon that was slain by the hero Hercules in Greek mythology, winds its way around the north celestial pole, all but enclosing the smaller Ursa Minor. Yet despite its size, Draco is disappointingly lacking in deep-sky objects and bright stars.

In mythology, Draco was the dragon that guarded the golden apples in the orchard of the Hesperides, and was fought by the great hero Hercules as one of his 12 labours. In today's skies, Hercules still kneels above the dragon's head, club raised to deliver a killing blow.

The constellation's brightest star is Eltanin, or Gamma Draconis, an orange giant some 150 light years from Earth that shines at magnitude 2.2. It happens to be heading towards Earth at a speed of around 28 kilometres per second (17.5 miles per second), and will pass within 28 light years of Earth around 1.5 million years from now, becoming the brightest star in the sky. Magnitude-3.6 Alpha Draconis, or Thuban, meanwhile, is a tight binary that, thanks to the effects of precession (see page 11) was pole star around 2,800 BC.

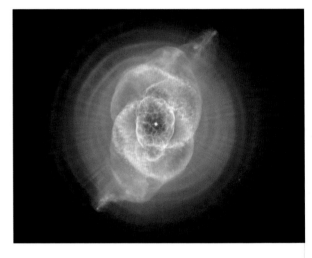

Cat's Eye Nebula
Draco's most celebrated object is the planetary nebula NGC 6543, also known as the Cat's Eye Nebula. Lying some 3,600 light years from Earth and faintly visible through a small telescope, the Cat's Eye is formed from glowing shells of gas cast by a giant star passing through a period of instability at the end of its life. This Hubble Space Telescope image reveals intricate spiral structures created as stellar winds blowing out from the central star catch up with and 'inflate' material cast off centuries before.

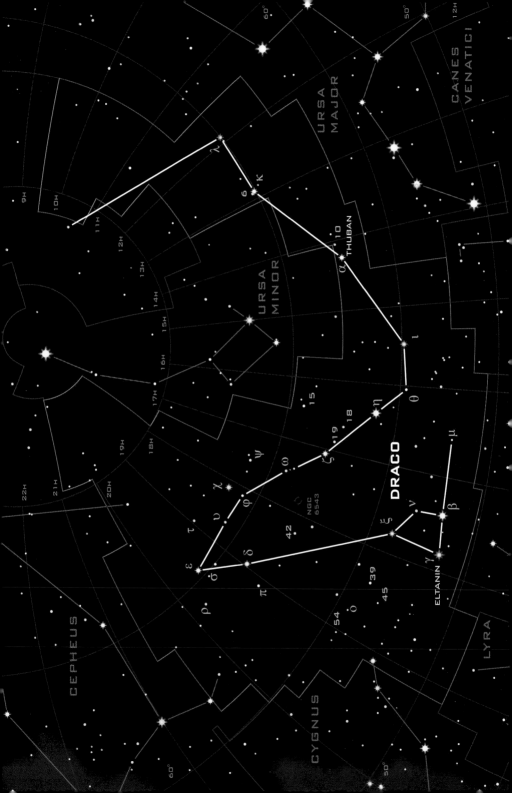

Cepheus

AT A GLANCE

NAME Cepheus
MEANING King Cepheus
ABBREVIATION Cep
GENITIVE Cephei
R.A. 02h 33m
DEC. +71° 01'
AREA 588 (27)
BRIGHTEST STAR Alderamin (α)

The brightest stars of this high northern constellation form a pattern resembling a child's drawing of a house, and are easily ignored among their more obvious and prominent neighbours. Yet with a position on the borders of the northern Milky Way, Cepheus has several overlooked treasures.

Cepheus represents the King of Ethiopia – husband of Cassiopeia and father of princess Andromeda in the Perseus legend (*see* page 62). The constellation's main claims to fame are its variable stars – the prototypes for no less than three different forms of pulsating stars are found here. Beta Cephei (Alfirk) is the brightest of these, a blue-white giant star some 595 light years from Earth whose variations of roughly 0.1 magnitude every 4.6 hours are too small and rapid a fluctuation to be seen without specialist equipment. Delta Cephei, in contrast, varies between magnitudes 3.5 and 4.4 every 5.37 days. Its changes are easily spotted by comparison with neighbouring stars, and it is the prototype for the class commonly known as 'Cepheids', which astronomers use to help them measure the distance of other galaxies.

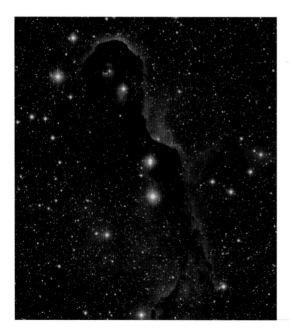

Garnet Star Nebula

A large nebula catalogued as IC 1396 lies close to the Garnet Star Mu Cephei, one of the most celebrated stars in the sky. Shining at magnitude 4.0 across 5,000 light years, Mu is a red supergiant, emitting as much energy as 350,000 Suns and varying unpredictably in brightness. IC 1396, some 2,000 light years closer to Earth, is rich in detail such as the dark dust lanes shown here.

Camelopardalis

AT A GLANCE

NAME Camelopardalis
MEANING The giraffe
ABBREVIATION Cam
GENITIVE Camelopardalis
R.A. 08h 51m
DEC. +69° 23'
AREA 757 (18)
BRIGHTEST STAR Beta (β)

This indistinct constellation, covering a large swathe of otherwise vacant sky, is a comparatively late addition to the northern heavens. It was only invented and named by a Dutch astronomer in the 17th century. It supposedly represents a giraffe, although it is hard to identify any real pattern among its faint stars.

Camelopardalis was introduced to celestial charts by the Dutch astronomer and theologian Petrus Plancius around 1600. Plancius apparently intended it to represent the camel that carried Rebecca to her wedding with Isaac in the Bible's Old Testament, but he mistakenly used the Latin word for a giraffe, and the name stuck.

The constellation's main objects of interest are two unpredictable variable stars, Beta and Z Camelopardalis. Beta is a yellow supergiant of magnitude 4.0 that is prone to unexpected surface flares: in 1967 it jumped in brightness by a whole magnitude for just a few minutes. Z Cam is a 'dwarf nova' – a double star system, normally of magnitude 13, which increases its brightness by roughly two magnitudes every few weeks, thanks to enormous explosions in the atmosphere of one of the stars.

Galaxy NGC 2403
The brightest galaxy in Camelopardalis lies 12 million light years away and shines at magnitude 8.4. However, because it is almost face-on to Earth its light is spread out, and it is best seen in dark skies using binoculars or a low-powered telescope. NGC 2403 forms part of the same small galaxy group as the better known Messier 81 and 82, over the border in Ursa Major (see page 32).

Cassiopeia

This distinctive W-shaped constellation is a fixture of far northern skies, circling around the north celestial pole opposite the equally distinctive Plough or Big Dipper. Cassiopeia represents the queen of the same name in the Perseus legend, the famously vain wife of King Cepheus and mother of the princess Andromeda.

AT A GLANCE

NAME Cassiopeia
MEANING Queen Cassiopeia
ABBREVIATION Cas
GENITIVE Cassiopeiae
R.A. 01h 19m
DEC. +62° 11'
AREA 598 (25)
BRIGHTEST STAR Shedar (α)

The constellation wheels high overhead on northern summer nights, and is home to the rich starfields of the northernmost Milky Way. Deep-sky objects include the open clusters M52, M103 and NGC 457, and the supernova remnant Cassiopeia A.

Alpha Cassiopeiae, or Schedar, is a yellow-orange giant star, 230 light years from Earth and 500 times more luminous than the Sun. Nineteenth-century astronomers considered it to be a variable star, but today it shines at a steady magnitude 2.25. Iota Cassiopeiae, meanwhile, is a multiple star system – with a magnitude-4.5 primary star, a magnitude-8.4 companion visible through small telescopes and a further component of magnitude 6.9, that is only resolvable through larger instruments.

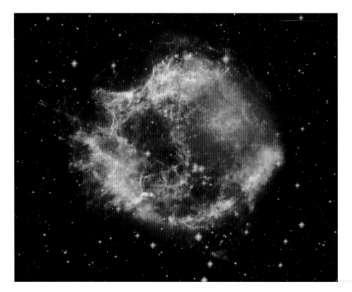

Cassiopeia A
A tattered, expanding bubble of superhot gas, barely visible at optical wavelengths but a strong source of both X-rays and radio waves, marks the site of the Milky Way galaxy's most recent known supernova explosion. This particular supernova, marking the death of a massive star 11,000 light years away, should have been seen on Earth around 300 years ago, but apparently went unnoticed at the time.

Auriga and Lynx

Auriga is a prominent constellation in mid-northern skies, dominated by the brilliant star Capella, which is the sixth brightest star in the night sky. This constellation is home to several interesting stars and clusters. Its near neighbour Lynx, in contrast, is renowned for its faintness and obscurity.

AT A GLANCE

NAME Auriga/Lynx
MEANING The charioteer/The lynx
ABBREVIATION Aur/Lyn
GENITIVE Aurigae/Lyncis
R.A. o6h o4m/o7h 6om
DEC. +42° o2'/+47° 28'
AREA 657 (21)/545 (28)
BRIGHTEST STAR Capella (α)/Alpha (α)

Although Auriga has been seen as a charioteer since ancient times, its precise identity varies – the figure may be one of two legendary Greek heroes, Eritchthonius or Myrtilus. The constellation's most prominent stars, meanwhile, are associated with a completely different story – Capella's name means 'she goat', and the star represents Amaltheia, the goat that suckled the infant Zeus. The tight triangle of stars just to the southwest of Capella, meanwhile, are known as 'the kids'. Capella itself is a complex quadruple star system at a distance of 42 light years from Earth, though it takes a moderate telescope to spot even the brightest of the primary star's companions.

In stark contrast, Lynx is a 17th-century invention of Polish astronomer Johannes Hevelius, who filled a gap in the sky with a chain of stars so faint, he said, that only the lynx-eyed could spot it.

AE Aurigae

At the limits of naked-eye visibility, AE Aurigae is a seemingly unremarkable blue star around 1,400 light years from Earth. Its radiation illuminates the faint but beautiful structure of IC 405, the so-called Flaming Star Nebula, but the star did not originate in this region of space – instead it is a rare, fast-moving runaway, ejected from its birthplace in the Orion Nebula around 2 million years ago.

Auriga and Lynx Inside View

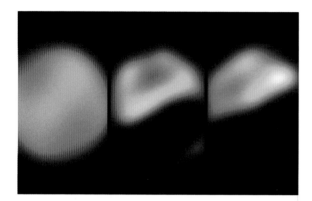

Epsilon (Ɛ) Aurigae

The northernmost of 'the kids' triangle of stars, Epsilon Aurigae is a remarkable variable star whose true nature has only recently been established. It normally shines at a steady magnitude 3.0, but every 27 years its brightness drops gradually to magnitude 3.8, where it remains for about a year before recovering. This behaviour naturally suggests an eclipsing binary star, but the eclipsing secondary object seems to be semitransparent and makes no contribution to the system's light at other times. What's more, the length of the eclipses suggests it must be around 3 billion km (1.9 billion miles) across. One long-standing theory to explain this strange behaviour is that the eclipse is caused by a disc of planet-forming material in orbit around an unseen object that is itself orbiting the primary star. This was finally confirmed during the 2009–11 eclipse, when astronomers imaged the disc crossing the face of Epsilon Aurigae for the first time.

R.A. 05h 02m, DEC. +43º49'
MAGNITUDE 2.9–3.8 (var)
DISTANCE 2,000 light years

The Intergalactic Wanderer NGC 2419

This faint globular cluster, only visible through medium-sized telescopes, stands out from others of its type for two main reasons. Firstly, it lies on the opposite side of the sky from most other globular clusters that orbit around the core of the Milky Way – there are no other globulars within 60 degrees of it. Secondly, and even more remarkably, it is almost 300,000 light years from Earth – much further away than our galaxy's other globular clusters, and even more distant than some of the Milky Way's satellite galaxies. It is currently approaching us at a speed of 20 km per second (12.5 miles per second), but astronomers suspect it is not gravitationally bound to the Milky Way at all – it may be an outcast from another galaxy altogether, flung off into intergalactic space during a cosmic close encounter.

R.A. 07h 38m, DEC. +38º53'
MAGNITUDE 10.4
DISTANCE 295,000 light years

IC 405 Flaming Star Nebula

This beautiful nebula around the speeding runaway star AE Aurigae is illuminated by a mixture of emission and reflection processes. Visible light from the central star is reflected and scattered by fine dust particles within the nebula, which tend to bend shorter wavelengths more than longer ones and therefore give the light deflected towards Earth a blue tint. Meanwhile, invisible ultraviolet radiation energizes the nebula's gas atoms and molecules, causing them to emit radiation at visible wavelengths as they return to their normal state.

R.A. 05h 16m, DEC. +34º27'
MAGNITUDE c.6.0 (var)
DISTANCE 1,400 light years

Ursa Major

AT A GLANCE

NAME Ursa Major
MEANING The great bear
ABBREVIATION UMa
GENITIVE Ursae Majoris
R.A. 11h 19m
DEC. +50° 43'
AREA 1,280 (3)
BRIGHTEST STAR Alioth (ε)

While the Plough or Big Dipper is the most instantly recognizable of star patterns for northern hemisphere observers, the constellation of the Great Bear extends across a much larger region. Homer mentions it in Book XVIII of the *Iliad* referring to 'the Bear, which men also call the Wain'.

Ursa Major is the third-largest constellation in the entire sky, and has been imagined as a bear by widely separated civilizations since ancient times. Its seven brightest stars form the body and tail of the bear, while fainter outlying stars mark the head and limbs. The Plough or Big Dipper is a famously useful celestial signpost – its two brightest stars Dubhe and Merak (on the front of the dipper's 'pan') point directly towards the pole star Polaris, while the arc of its tail can be extended past Arcturus in Boötes, as far as Spica in Virgo. Many of the constellation's stars are physically close to each other in space, around 80 light years from Earth, and moving in the same direction through space – indications that they all originated in the same star cluster. Among the Plough stars, only Alpha Ursae Majoris (Dubhe) and Eta (Alkaid) are not members of the so-called 'Ursa Major Moving Group'.

Pinwheel Galaxy
The large face-on spiral galaxy Messier 101 forms a rough triangle with Mizar and Alkaid, the two easternmost stars of the Big Dipper's handle. Although ostensibly quite bright at magnitude 7.9, its light is diffuse because it happens to lie face-on as seen from Earth. As a result, it is best observed through binoculars or a telescope at low power, where it appears as a pale patch of light about half the size of the Full Moon. The Pinwheel lies some 27 million light years away.

Ursa Major Inside View

Mizar and Alcor
ζ Ursae Majoris

The middle star of the Bear's tail, Zeta Ursae Majoris or Mizar, is a renowned multiple star. Binoculars or even sharp eyesight reveal a magnitude-4.0 companion, Alcor, close to magnitude-2.3 Mizar, and even a small telescope will show that Mizar is itself binary. What's more, each of these three stars is a double star in turn. For a long time, the grouping of Mizar and Alcor was seen as mere chance, but new studies have suggested that the stars are genuinely bound by gravity, creating a sextuple system.

R.A. 13h 25m, DEC. +54°56'
MAGNITUDE 2.3, 4.0
DISTANCE 78 light years

Hubble Deep Field

Over ten days in December 1995, astronomers used the Hubble Space Telescope to take a series of images of a small patch of apparently empty sky in Ursa Major. Combining the multiple exposures using computerized image processing, NASA scientists produced the Hubble Deep Field (HDF) – a view that reveals some 3,000 galaxies stretching across billions of light years of space and offers a new perspective on the Universe.

R.A. 12h 37m, DEC. +62°12'
MAGNITUDE < 28
DISTANCE Up to 12 billion light years

Bode's Galaxy
Messier 81

This tightly wound spiral galaxy, 12 million light years from Earth, bears the name of Johann Elert Bode, the German astronomer who discovered it in 1774. Binoculars show it as a fuzzy point of light while a small telescope will reveal the oval shape of its central nucleus and larger instruments may trace its spiral arms. Although most of the light from its central regions is due to sheer density of stars, M81 also emits some radiation directly from its nucleus, making it one of the closest 'active galaxies'. It forms the core of one of the closest galaxy groups to our own.

R.A. 09h 56m, DEC. +69°04'
MAGNITUDE 6.9
DISTANCE 12 million light years

Ursa Major The Cigar Galaxy M82

An Exploding Galaxy

A popular target for amateur stargazers, the Cigar Galaxy is one of our brightest galactic neighbours, with a compact structure that shines at magnitude 8.4. Its elongated shape, curious appearance and bright central regions led to its classification as an irregular galaxy, but in 2005 astronomers discovered traces of a spiral structure. This Hubble Space Telescope image combines light taken with four filters including visible and infrared light, and shows the emissions from filaments of hydrogen that surround the galaxy and give rise to its explosive appearance.

R.A. 09h 56m, DEC. +69°41'
MAGNITUDE 8.4
DISTANCE 12 million light years

Multiwavelength Composite

This colourful image combines data from visible-light, infrared and X-ray telescopes to produce a unique view of the Cigar Galaxy. Infrared observations from the Spitzer Space Telescope are shown in red, a visible-light image from the Hubble Space Telescope in white, and X-ray data from the Chandra X-ray Observatory in blue. The composite shows how X-ray emissions are concentrated in plumes that emerge from above and below the galaxy's nucleus – they are mostly 'synchrotron radiation' generated by charged electron particles moving at high speeds after their expulsion from the galaxy's active core.

Starburst Core

A detailed Hubble view of the Cigar's central regions reveals brilliant, compact star clusters embedded in a dusty environment. Each of these 'super star clusters' contains 100,000 stars or more – they are thought to be the infant stage of the globular clusters found in orbit around our own galaxy. Their formation is triggered by violent galactic encounters, in this case a glancing collision between M82 and the nearby spiral Messier 81 (*see* page 35). The properties of stars within the clusters suggest that they were created in a 100-million-year 'starburst' that began around 600 million years ago. The repercussions of this event are still being felt as disrupted material continues to fall onto the galaxy's central supermassive black hole, fuelling its ongoing activity.

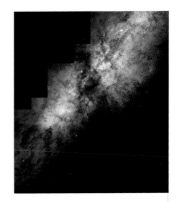

Canes Venatici

AT A GLANCE

NAME Canes Venatici
MEANING The hunting dogs
ABBREVIATION CVn
GENITIVE Canun Venaticorum
R.A. 13h 07m
DEC. +40° 06'
AREA 465 (38)
BRIGHTEST STAR Cor Caroli (α)

Although it has only one bright star to its name, the constellation of the Hunting Dogs is easily located thanks to its position between the familiar Plough or Big Dipper and the bright star Arcturus in Boötes. Its apparent emptiness belies the presence of several interesting objects.

Canes Venatici was added to the sky by Polish astronomer Johannes Hevelius in the late 1600s – prior to this time Arab astronomers had associated the area with the crook of the shepherd Boötes. The brightest star, Cor Caroli or 'Charles's Heart', was named by British astronomer Edmond Halley to commemorate the executed King Charles I who was beheaded in 1649. At magnitude 2.9, it is an attractive double, easily split with a small telescope to reveal a companion of magnitude 5.6. A little way to the north of Beta lies Y CVn, or 'La Superba', a red giant star on the verge of transforming into a planetary nebula. It fluctuates between magnitudes 4.8 and 6.3 in around 158 days. Another highlight of this small constellation is the globular cluster M3. At 34,000 light years from Earth, this rich star cluster shines at magnitude 6.2.

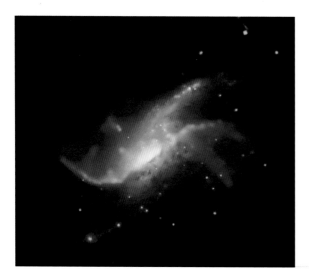

Messier 106
Tucked into the northwestern corner of Canes Venatici, M106 is an unusual spiral galaxy around 25 million light years from Earth. It was discovered in 1781 and is today classified as a Seyfert II galaxy – a type of active galaxy that has an unusually bright central region emitting radiation at a variety of wavelengths. Its strangest features are 'anomalous' arms that are invisible in normal light (shown blue and purple in this composite image).

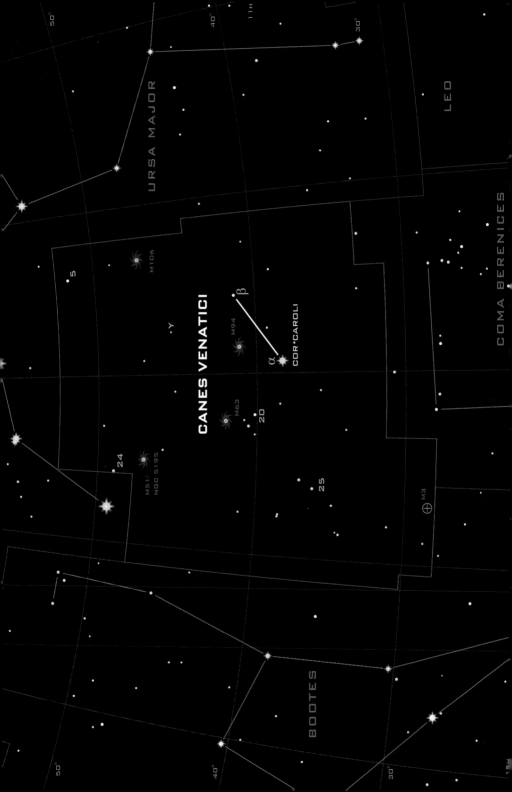

Canes Venatici
The Whirlpool Galaxy M51

Infrared Skeleton

This pair of Hubble Space Telescope images contrast the Whirlpool in visible light (left) and near-infrared radiation (right). Most of the galaxy's stars emit their light at visible wavelengths, so the infrared image effectively filters them out, revealing instead the distribution of warm gas and dust too cool to emit visible light. A comparison reveals that, while the disc of the galaxy is filled with this material, the opaque dust lanes that follow the Whirlpool's spiral structure are distinctly warmer than their surroundings. This is because they are the key locations where new stars are beginning to form, but have not yet become hot enough to shine properly. Interestingly, the galaxy's bright hub shows no clear difference from its surroundings – ongoing star formation in spirals such as the Whirlpool and our own galaxy is concentrated in the spiral arms rather than the central regions.

Bright Nucleus

This early image from the Hubble Space Tele
star-like point of light at the centre of the ga
reveals the presence of a dark cross in front
and for some time this was thought to indica
encircling a central supermassive black hole
have now shown that the cross is due to light
the foreground. However, the brightness of t
indicates the presence of a black hole – as do
emitting lobes of multi-million-degree gas, v
hundred light years on either side of the nuc

Dancing Partners

At magnitude 8.4, Messier 51 is one of the brightest galaxies in the sky despite an impressive distance of 23 million light years. Binoculars can locate its diffuse form, but a medium-sized telescope is needed to pick out its spiral arms. One of M51's spiral arms appears to link it to the nearby irregular galaxy NGC 5195, but this is a line-of-sight effect (in fact NGC 5195 lies behind the spiral arm).

However, the galaxies are still close enough to affect each other – M51's gravity may have triggered a burst of star formation in its smaller neighbour, while NGC 5195's influence may have awoken M51's active nucleus and helped to reinforce its spiral pattern.

R.A. 13h 30m, DEC. +47°12'
MAGNITUDE 8.4
DISTANCE 23 million light years

Boötes

AT A GLANCE

NAME Boötes
MEANING The herdsman
ABBREVIATION Boo
GENITIVE Bootis
R.A. 14h 43m
DEC. +31° 12'
AREA 907 (13)
BRIGHTEST STAR Arcturus (α)

This constellation's distinctive kite-like shape and brightest star Arcturus make it a familiar feature of northern skies as it chases the Greater and Lesser Bears around the celestial pole. The radiant of the Quadrantid meteor shower, which occurs in January, is located in northern Boötes near Kappa Boötis.

Boötes is traditionally said to represent a celestial herdsman, driving the bears away from his flock with the assistance of his hunting dogs (nearby Canes Venatici). In another interpretation, however, he represents Arcas, son of Zeus and the beautiful nymph Callisto, who was transformed into the Great Bear by the jealous goddess Artemis, who was herself a daughter of Zeus.

With a name meaning 'bear follower', the constellation's brightest star, Arcturus, is the closest red giant star to Earth, just 37 light years away. It is easily located by following the arc of the three stars in the handle of the Big Dipper, and lies at the base of an elongated kite shape that forms the body of the constellation. Arcturus is thought to be a star quite similar to the Sun, but more advanced along its life cycle. Having exhausted the hydrogen fuel supply at its core, it is now advancing towards the final stages of its evolution.

Tau Boötis
This Sun-like star, some 51 light years from Earth, is home to one of the first extrasolar planets to be discovered. Tau Boötis b has roughly four times the mass of Jupiter, but orbits its star in just 3 days 7.5 hours. Much further out, a red dwarf binary companion circles the primary star in a thousand-year orbit. Discovered in 1996, Tau Boötis b was one of the first extrasolar planets in a class known as the 'hot Jupiters'.

Corona Borealis

This distinctive circlet of stars, lying close to Boötes in a fairly empty region of the northern sky, makes up for a lack of deep-sky objects with interesting variable stars. In Greek mythology this constellation represents the crown that the god Dionysus presented to his bride Ariadne on the isle of Naxos.

AT A GLANCE

NAME Corona Borealis
MEANING The northern crown
ABBREVIATION CrB
GENITIVE Coronae Borealis
R.A. 15h 51m
DEC. +32° 37'
AREA 179 (73)
BRIGHTEST STAR Alphecca (α)

The Northern Crown is one of the 48 constellations listed by the Greek-Egyptian astronomer Ptolemy in the second century AD. R Corona Borealis, near the centre of the coronet, is a yellow supergiant 6,000 light years from Earth. Normally it shines at magnitude 5.9, on the limit of naked-eye visibility, but it sometimes drops to magnitude 14, beyond the reach of most amateur telescopes. Astronomers think these unpredictable plunges are triggered when the star throws large amounts of carbon into its atmosphere, where it forms dark clouds. Meanwhile, 1,800 light years away and close to Epsilon, the 'Blaze Star' T Corona Borealis shows the reverse behaviour. It is a 'recurrent nova' system, normally languishing at around magnitude 11.0 but prone to huge explosions that can raise it to magnitude 2.0 every few decades.

Abell 2065

This dense galaxy cluster lies in the southwest corner of Corona Borealis, and is one of the most distant visible with amateur equipment. Despite this, its combined magnitude of around 14.0 still requires a fairly large amateur telescope or a long photographic exposure. Abell 2065 is one of many distant clusters catalogued by US astronomer George Abell in the 1950s. It contains more than 400 galaxies and lies around a billion light years from Earth.

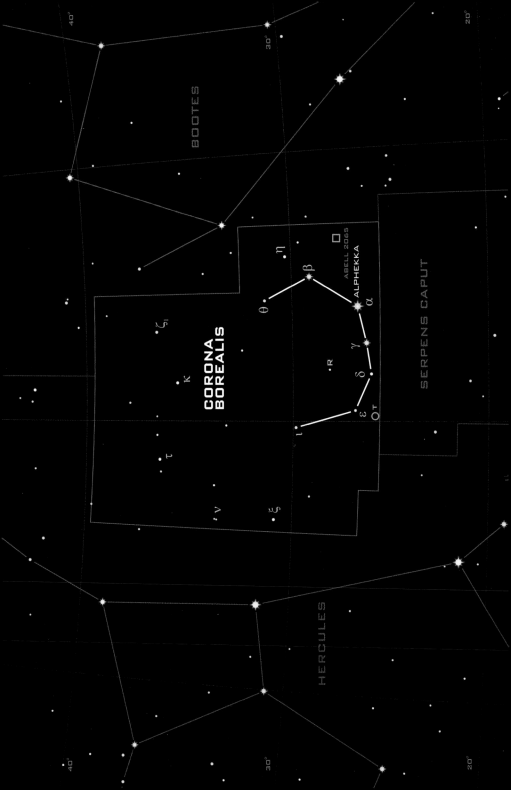

Hercules

AT A GLANCE

NAME Hercules
MEANING Hercules
ABBREVIATION Her
GENITIVE Herculis
R.A. 17h 23m
DEC. +27° 30'
AREA 1,225 (5)
BRIGHTEST STAR Kornephoros (β)

This large northern constellation represents the great hero of Greek and Roman myth. Earlier Greek astronomers knew it as Engonasin, or 'the kneeling man'. Despite its size, Hercules lacks in bright stars – its most prominent deep-sky objects are globular star clusters, including one of the sky's finest.

Best located by looking midway between the bright stars Vega in Lyra and Arcturus in Boötes, Hercules consists of a lopsided quadrangle of stars known as the Keystone, with extended chains of stars at each corner marking the hero's limbs. He is usually depicted upside down, kneeling with his foot on the head of the dragon Draco, and a club in one hand. In Greek myth, he was famously forced to accomplish 12 tasks to atone for killing his family. The hero's head is represented by the double star Rasalgethi, which is easily resolved through a small telescope to reveal a bright red star of magnitude 4.5 and a greenish-white companion of magnitude 5.4. Delta Herculis is another attractive double for small telescopes, consisting of a blue primary of a magnitude 3.1, with a companion of magnitude 8.2.

Messier 13

Hercules is home to the finest globular cluster in the northern sky – Messier 13. This huge cluster contains roughly a million stars crammed into a ball just 150 light years across. At a distance of 25,000 light years from Earth, it is just visible to the naked eye, and appears through binoculars as a circular patch of light. Small telescopes can pick out individual chains of stars around its outer edges.

Lyra

AT A GLANCE

NAME Lyra
MEANING The lyre
ABBREVIATION Lyr
GENITIVE Lyrae
R.A. 18h 51m
DEC. +36° 41'
AREA 286 (52)
BRIGHTEST STAR Vega (α)

This compact constellation, instantly recognizable thanks to the presence of Vega, the fifth-brightest star in the sky, represents the ancient musical stringed instrument – looking rather like a small harp – played by the Greek hero Orpheus. His music was so beautiful that even trees and rocks would dance to it.

Lying close to the dense star clouds of the northern Milky Way, Lyra is home to several interesting objects, including the famous Ring Nebula.

Vega, or Alpha Lyrae, is not simply interesting on account of its brightness. It is comparatively close – just 25 light years from Earth – and quite young at about 500 million years old (one-tenth of the age of the Sun). Infrared studies have revealed that it is still surrounded by a disc of gas and dust that may be in the process of forming planets.

Another highlight of the constellation is Epsilon Lyrae, a famous multiple star system. Binoculars show that it consists of two distinct components of magnitudes 4.7 and 4.6, but small telescopes will reveal that each of these stars is a double in its own right. Beta Lyrae or Sheliak, meanwhile, is a much tighter system – an 'eclipsing binary' star.

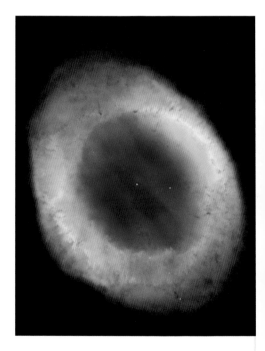

Ring Nebula
Midway between Beta and Gamma Lyrae lies Messier 57, a faint cosmic smoke ring best seen through a small telescope. The Ring is a famous planetary nebula – a complex expanding shell of gas shrugged off by a dying Sun-like star, and energized by radiation from the hot star at its centre. It lies around 2,300 light years from Earth and is about 1.5 light years across.

Vulpecula and Sagitta

AT A GLANCE

NAME Vulpecula/Sagitta
MEANING The fox/The arrow
ABBREVIATION Vul/Sge
GENITIVE Vulpeculae/Sagittae
R.A. 20h 14m/19h 39m
DEC. +24° 27'/+18° 52'
AREA 268 (55)/80 (86)
BRIGHTEST STAR Anser (α)/Gamma (γ)

Tucked alongside the more distinctive constellations of Cygnus and Lyra, these two small and obscure star groups nevertheless contain some interesting objects. The arrow shape of Sagitta, pointed towards the celestial birds Cygnus and Aquila, is the more easily recognized of the two constellations.

Sagitta's brightest star, misleadingly designated Gamma, is one of the few cool M-class stars visible to the naked eye – a dying red giant about 275 light years away. It is also home to Messier 71, a loosely bound globular cluster 13,000 light years from Earth.

The shapeless and faint Vulpecula, meanwhile, supposedly represents a fox running with a goose in its jaws. The goose is supposedly marked by the constellation's brightest star, known as Anser. Other highlights include the Dumbbell Nebula, and the distinctive coathanger-shaped group of stars known as Brocchi's Cluster, or Collinder 399. In 1967, astronomers surveying the constellation with radio telescopes, discovered PSR B1919+21, a steadily flashing radio source that proved to be the first known pulsar (a rapidly rotating neutron star, see page 76).

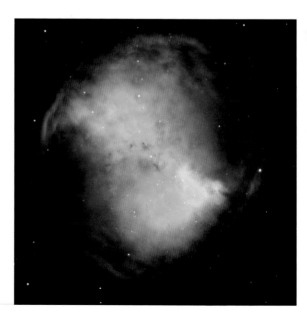

Dumbbell Nebula
Also known as Messier 27, the Dumbbell was the first planetary nebula to be discovered, by French astronomer Charles Messier in 1764. At 1,350 light years away, it is one of the closest objects of its kind to Earth, and is quite easy to spot through binoculars as a patch of light one-third the size of the Full Moon.

Cygnus

AT A GLANCE

NAME Cygnus
MEANING The swan
ABBREVIATION Cyg
GENITIVE Cygni
R.A. 20h 35m
DEC. +44° 33'
AREA 804 (16)
BRIGHTEST STAR Deneb (α)

Often called the Northern Cross, the bright stars of Cygnus form a prominent pattern in northern skies, encompassing some of the richest starfields in the Milky Way and a variety of fascinating deep-sky objects. The brightest star in Cygnus is Deneb. Along with Vega and Altair, it forms the Summer Triangle.

The ancient Greeks viewed Cygnus as one of several mythological swans, flying along the Milky Way. To some, it represented Zeus, the king of the gods himself, taking the form of a swan in order to seduce the beautiful Leda. To others, it was the musician Orpheus, transformed into a swan after his death and placed in the sky alongside his lyre. A third legend saw Cygnus as the loyal friend of Phaeton, a reckless youth who stole the Sun god's chariot and crashed it into the celestial river Eridanus. According to this interpretation, Cygnus repeatedly plunged into the river in a vain attempt to save his friend, and so for his valour the god Zeus preserved him as a swan among the stars of the heavens.

Whatever its identity, Cygnus is an astronomical treasure trove. Its brightest star, Deneb, is the most intrinsically luminous of all first-magnitude stars, shining with the brilliance of 200,000 Suns and still appearing brilliant across a distance of 3,000 light years.

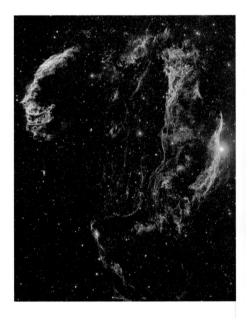

Veil Nebula
One of the brightest supernova remnants in the sky, the diaphanous Veil Nebula covers an area of the heavens roughly six times the diameter of the Full Moon, a little way to the southeast of Epsilon Cygni. These shredded sheets of hot gas have now expanded into a shell around 50 light years across. The brightest segments, discovered by the British astronomer William Herschel in 1784, have separate NGC catalogue numbers 6960, 6992 and 6995.

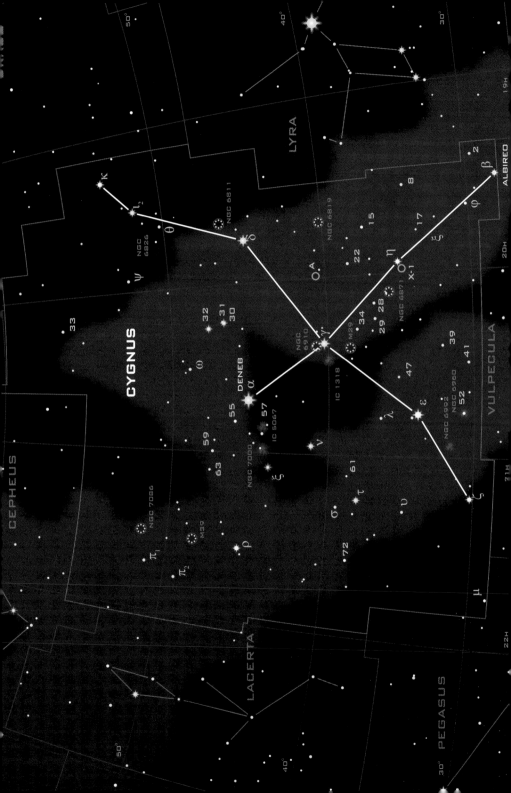

Cygnus Inside View

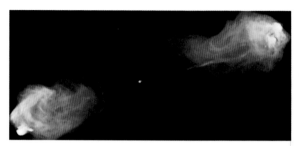

Cygnus X-1

Visible-light observations of this strange object reveal a brilliant blue supergiant, reduced to a feeble 9th magnitude by its distance of more than 8,000 light years. However, X-ray studies reveal something else entirely – a strong source of high-energy X-rays flickering a thousand times per second. The rays do not come directly from the star, but from an otherwise invisible object that orbits it once every 5.6 days and weighs as much as five Suns. This suggests that the secondary object is probably a black hole, pulling gas away from the visible star and forming a superhot disc of X-ray-emitting material that spirals inwards to its doom.

R.A. 19h 58m, DEC. +35°12'
MAGNITUDE 8.9
DISTANCE 8,200 light years

Cygnus A

In visible light, this distant and distorted galaxy can only be detected through large telescopes, but for radio astronomers it is one of the brightest objects in the sky. A pair of tight jets emerge on opposite sides of the visible galaxy and billow out to form a huge double-lobed structure of radio-emitting gas half a million light years across. Cygnus A is one of the nearest and most powerful radio galaxies – a type of active galaxy in which the central supermassive black hole 'engine' is hidden from view so that all we see are the effects of material ejected from above and below this central region.

R.A. 19h 59m, DEC. +40°44'
MAGNITUDE 15.0
DISTANCE 600 million light years

Albireo
β Cygni

Arguably the most beautiful double star in northern skies, Albireo is an observing treat through even the smallest telescope. The system comprises a yellow-orange star of magnitude 3.1 and a contrasting blue-green star of magnitude 5.1. Both stars lie around 385 light years from Earth, but astronomers have not confirmed that they are actually in orbit around each other. In 1976 astronomers confirmed that the brighter component, Albireo A, is itself binary, but its two elements are all but impossible to separate visually, even with professional equipment.

R.A. 19h 31m, DEC. +27°58'
MAGNITUDE 3.1/5.1
DISTANCE 385 light years

The Witch's Broom NGC 6960

The brightest section of the Veil Nebula, marking its western edge, is sometimes known as the Witch's Broom. Here the nebula's light is relatively concentrated, and the naked-eye star 52 Cygni (magnitude 5.3) acts as a useful pointer. This segment of the Veil is roughly 1.5 degrees across – three times the size of the Full Moon. At an estimated distance of almost 1,500 light years, meanwhile, the entire Veil complex covers a volume of space 50 light years across.

R.A. 20h 46m, **DEC.** +30°43'
MAGNITUDE 7.0
DISTANCE 1,470 **light years**

Cygnus North America Nebula NGC 7000

Continent in Space

The celebrated North America Nebula covers an area of sky four times the size of the Full Moon, east of the constellation's brightest star Deneb. The resemblance to the continent of North America is clear, with a dark dust lane forming the Atlantic coast and Gulf of Mexico. On the other side of this dust lane lies another distinctive shape, the Pelican Nebula IC 5070. Although NGC 7000's 'integrated magnitude' is estimated at around 4.0, the size of the nebula means that its light is spread across a large area, so it is hard to pick out from the background Milky Way with the naked eye. Instead, the nebula is best observed under dark, moonless skies with binoculars or a telescope with low magnification and a wide field of view.

Ionization Front

These colourful peaks are found within the Pelican Nebula, separated from the North America Nebula proper by a canyon of opaque dust but essentially part of the same star-forming complex. The mountainous outcrops have formed as radiation from newborn stars concentrated along the canyon blows away the surrounding gas. Radiation absorbed and then re-emitted by the gas causes it to glow in colours that are linked to its composition – while some gas is ionized (transformed into electrically charged particles), creating the characteristic blue haze above the peaks themselves. Denser areas within the nebula hold out against the torrent of radiation for longer. For instance, the tendril-like structures at the top of each peak form where relatively dense knots of material cast a protective 'shadow'. These knots may ultimately condense to create further young stars.

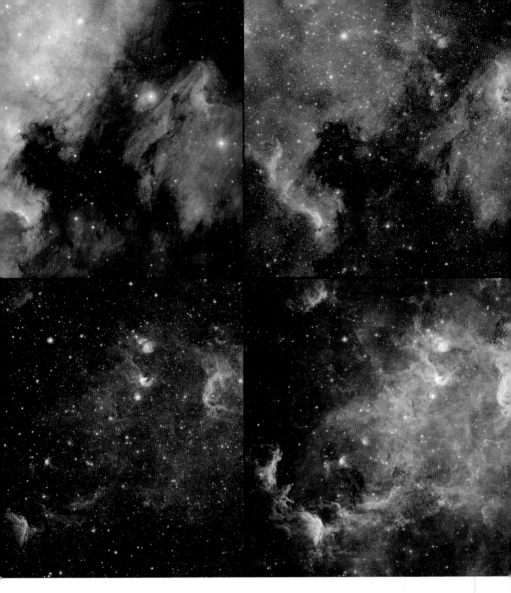

Multiwavelength Montage

A series of images reveals the inner regions of the North America Nebula in four different wavelengths. At top left, a visible-light view emphasizes the similarity to North America, and also highlights the Pelican Nebula and intervening dust lane. At top right, visible light (blue) has been combined with cooler infrared radiation (red) to show the warm glow of the nebula's dust clouds. Mid-infrared (bottom left) and far-infrared (bottom right) images pierce the dust completely to reveal the glow of newborn stars embedded within the nebula.

R.A. 20h 59m, **DEC.** +44°20'
MAGNITUDE 4.0
DISTANCE 1,600 light years

Andromeda and Lacerta

The fairly indistinct branching shape of Andromeda is still easy to locate in the sky because its brightest star is also the northeast corner of the Square of Pegasus. The more compact and distinctive shape of Lacerta can be found to its northwest. It resembles a lizard as if viewed from overhead.

AT A GLANCE

NAME Andromeda/Lacerta
MEANING Princess Andromeda/
The lizard
ABBREVIATION And/Lac
GENITIVE Andromedae/Lacertae
R.A. 00h 48m/22h 28m
DEC. +37° 26'/+46° 03'
AREA 722 (19)/201 (68)
BRIGHTEST STAR Alpheratz (α)/Alpha (α)

Andromeda is an ancient constellation – in mythology, she was the beautiful daughter of Cepheus and Cassiopeia of Ethiopia. Cassiopeia's boasts about Andromeda enraged the jealous Juno, queen of the gods, who sent the sea monster Cetus to ravage their kingdom until Andromeda was sacrificed. Lacerta is a more recent invention, created by Johannes Hevelius around 1687.

Alpha Andromedae, or Alpheratz, is a slightly variable blue-white star of magnitude 2.1, while Gamma, or Almach, is an attractive multiple. A small telescope will split it into yellow and blue stars of magnitudes 2.3 and 4.8. The blue star has a fainter companion of magnitude 6.1 that is currently drawing away from its closest alignment in 2012, when the pair were indivisible through even the largest telescopes.

Blue Snowball

The planetary nebula NGC 7662, 2,200 light years from Earth, lies midway between Iota and Omicron Andromedae and is one of the easiest objects of its kind for amateurs to detect at magnitude 9.0. A small telescope will show it as a blue-green star-like point, while medium-sized instruments reveal a distinct disc.

Dust and Stars

This unusual view of Andromeda, captured by NASA's Wide-Field Infrared Survey Explorer (WISE) shows the galaxy in a variety of infrared wavelenths. Near-infrared radiation from comparatively hot objects is shown in blue, highlighting the locations of the galaxy's less massive, cooler and longer-lived stars. Longer wavelengths from cooler regions are shown in green and red – by highlighting regions such as starbirth nebulae, they also trace the locations where hotter, short-lived stars are found. These massive, hot stars only survive for a few million years so they tend to dominate young open clusters and rarely have time to migrate far from the site of their origin. As a result, the warmer yellow areas trace the galaxy's star-forming spiral arms, revealing a complex structure that may have been shaped by past collisions between galaxies.

Double Nucleus

This enhanced image of the galaxy's nucleus from the Hubble Space Telescope reveals two distinct concentrations of stars at the heart of Andromeda. The true centre of M31 coincides with the fainter of these two structures, and until recently astronomers had trouble explaining the galaxy's apparent 'double nucleus'. One popular theory suggested that the brighter nucleus was the remnant of another galaxy absorbed into Andromeda at some point in the past. However, improved computer models suggest that the nucleus of a second galaxy could not persist for long without losing its integrity, so the currently accepted theory is that the brighter concentration comes from a disc of stars orbiting the true centre of M31 and bunching up to form a stellar 'traffic jam' when they are at their furthest from the nucleus.

Island Universe

The magnificent Andromeda Galaxy is clearly visible with the naked eye under dark skies, as an elongated patch of light about twice the size of the Full Moon. Small telescopes will show the bright knot of the galaxy's nucleus floating in the fainter glow of the galaxy's broad disc, but larger instruments or long-exposure photographs are required to bring out features such as the dark dust lanes that help define the spiral structure. With a diameter of around 200,000 light years, M31 is far larger than our galaxy but seems to have less mass. Despite this, it forms the other major gravitational centre in our 'Local Group' of galaxies, and is powerful enough to attract satellite galaxies including M32 and M110, shown in this image.

R.A. 00h 43m, DEC. +41°16'
MAGNITUDE 3.4
DISTANCE 2.5 million light years

Perseus

AT A GLANCE

NAME Perseus
MEANING Perseus
ABBREVIATION Per
GENITIVE Persei
R.A. 03h 11m
DEC. +45° 01'
AREA 615 (24)
BRIGHTEST STAR Mirphak (α)

This constellation represents a hero whose legend unfolds across the northern sky. Andromeda, Perseus' wife in Greek myth, lies to the west and Cassiopeia, Andromeda's mother, to the north. However, its awkward shape and position against rich Milky Way starfields can make it hard to identify.

In Greek myth, Perseus was an exiled prince of Argos. With the help of the goddess Athena, he slew the gorgon Medusa, then went on to rescue Andromeda from the monster Cetus. He is usually depicted brandishing Medusa's head in one hand.

The constellation's most famous star, Beta Persei, or Algol, is said to mark the eye of Medusa's head brandished in the hero's hand. Algol normally shines at a steady magnitude 2.1, but dips abruptly to magnitude 3.4 for ten hours in a cycle that repeats every 2 days and 21 hours. It is the most prominent 'eclipsing binary' in the sky – a system in which two stars in close orbit around each other occasionally pass in front of one another, reducing the system's overall brightness. Algol's behaviour has been recognized since at least 1670, but its name, from the Arabic for 'demon', suggests its strange nature may have been known since ancient times.

California Nebula
This softly glowing nebula, catalogued as NGC 1499, lies close to the magnitude 4.0 star Xi Persei and is roughly five times the width of the Full Moon. It takes its name from its resemblance to the US state in long-exposure photographs. Xi Persei, at the centre of this image, is one of the most massive, hottest and most luminous stars in the sky, with the mass of 40 Suns and a total luminosity 330,000 times that of our star. It lies around 1,800 light years from Earth.

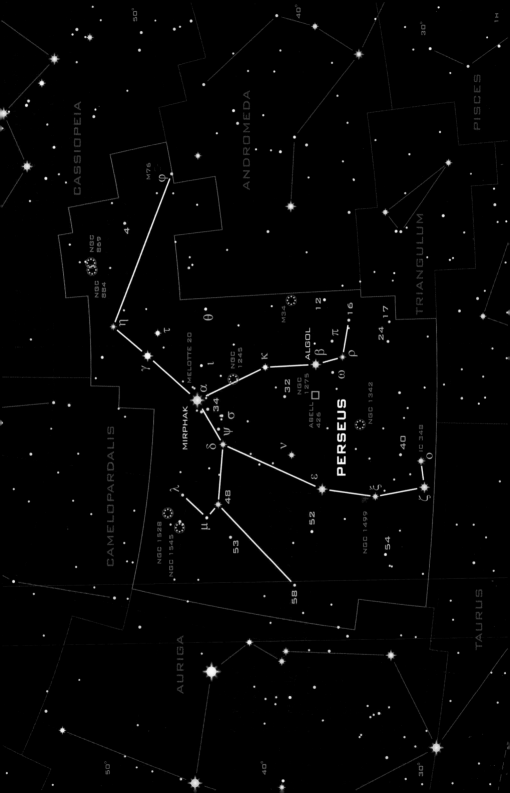

Perseus Inside View

Double Cluster
NGC 869 and 884

Midway between Perseus and Cassiopeia lies one of the most beautiful sights in northern skies. What looks to the naked eye like a pair of fuzzy 'stars', transforms in binoculars or a small telescope into a pair of rich open clusters studded with stars. The two clusters are separated by around 300 light years and so are not bound together by gravity, but they share a common origin in a group of stars and clusters known as the Perseus OB1 Association. Both clusters are relatively young – NGC 869 is around 19 million years old, and NGC 884 is 12.5 million years old – so they are dominated by a mix of massive, brilliant blue-white stars and even more massive red and orange giants that are already approaching the end of their lives.

R.A. 02h 21m, DEC. +57°08'
MAGNITUDE 4.3 and 4.4
DISTANCE 7,100 and 7,400 light years

Perseus Cluster
Abell 426

A little way to the east of Algol lies the heart of one of the most massive galaxy clusters in the local Universe, around 240 million light years from Earth. The Perseus Cluster dwarfs even the nearby Virgo Cluster (see page 92), containing as it does thousands of individual galaxies. The space around the cluster galaxies is filled with multi-million-degree, X-ray-emitting gas that has been stripped from the galaxies themselves during billions of years of collisions and close encounters. The giant galaxy NGC 1275 is embedded in the hottest part of this gas at the heart of the cluster.

R.A. 03h 18m, DEC. +41°30'
MAGNITUDE 12.6
DISTANCE 240 million light years

Perseus A

The giant galaxy at the centre of the Perseus Cluster, NGC 1275, coincides with a strong radio source known as Perseus A. Hubble Space Telescope images have revealed the galaxy's peculiar structure – most likely a result of a collision between two spiral galaxies. The collision has triggered a wave of activity in the nucleus of NGC 1275, pouring material into the supermassive black hole at its core, and triggering the emission of intense light and other radiations including radio waves. According to one popular theory, mergers between spirals within crowded galaxy clusters strip them of star-forming gas and eventually transform them into giant elliptical galaxies.

R.A. 03h 19m, DEC. +41°30'
MAGNITUDE 12.6
DISTANCE 237 million light years

Pisces

This relatively faint but distinctive zodiac constellation consists of two chains of stars lying to the east of the brighter stars of the Great Square of Pegasus. Each chain represents a fish, and they are often depicted tied together at their tails, with Alpha Piscium, or Alrescha, marking the cord that binds them.

AT A GLANCE

NAME Pisces
MEANING The fishes
ABBREVIATION Psc
GENITIVE Piscium
R.A. 00h 29m
DEC. +13° 41'
AREA 889 (14)
BRIGHTEST STAR Alrescha (α)

In ancient times, however, the fish were visualized as swimming freely, and were thought to represent Aphrodite and Cupid fleeing the approach of the sea monster Typhon (nearby Cetus). Alrescha is a good double star through a moderate telescope, while Zeta Piscium is another double that can be resolved into its components through even the smallest instrument.

The 'First Point of Aries', marking the northern vernal equinox where the ecliptic crosses from the southern to the northern hemisphere of the sky (*see* page 12) lies near the constellation's southern edge. It marks the origin point for the entire equatorial coordinate system that astronomers use for mapping the sky.

Messier 74
A little to the east of Eta Piscium, this spectacular spiral galaxy is best seen through larger telescopes at low magnification. Although relatively nearby at 30 million light years, it lies face-on to Earth, so its light is spread across the sky. Long-exposure photographs, however, reveal a beautiful geometric structure based around two spiral arms. Messier 74 is often cited as a 'Grand Design' spiral, in which the spiral arms, sites of new star formation, are very well defined.

Aries and Triangulum

AT A GLANCE

NAME Aries/Triangulum
MEANING The ram/The triangle
ABBREVIATION Ari/Tri
GENITIVE Arietis/Trianguli
R.A. 02h 38m/02h 11m
DEC. +20° 48'/+31° 29'
AREA 441 (39)/132 (78)
BRIGHTEST STAR Hamal (α)/Beta (β)

The relatively faint zodiac constellation of the Ram is somewhat lacking in interesting objects, although several meteor showers appear to radiate from Aries, including the Daytime Arietids. By contrast, a nearby compact triangle of faint stars offers the wonders of one of our closest galactic neighbours.

The hunched pattern of Aries has been identified as a crouching ram since well before the heyday of Greek astronomy, but today the constellation is usually seen as the ram that bore the golden fleece sought by Jason and the Argonauts. The wedge-shaped triangle of stars to its north was first noted by Greek astronomers on account of its similarity to the capital letter Delta (D).

Alpha Arietis, or Hamal, is an orange giant of magnitude 2.0, about 66 light years from Earth. It is a rare example of a star whose size has been directly measured, and is known to be 14.7 times bigger than the Sun. Gamma Arietis or Mesarthim, meanwhile, is an attractive multiple, 205 light years away – its twin white stars of magnitudes 4.6 and 4.7 are easily split by the smallest telescope, while the sharp-eyed may spot a smaller orange star of magnitude 9.6 that orbits around them at a greater distance.

Galactic Encounter
Northwest of Beta Arietis lies a pair of galaxies best seen through a medium-sized telescope. NGC 678 (at left in this photograph) is an edge-on spiral with a dark lane of dust across its centre. NGC 680, meanwhile, is an elliptical – a ball of stars lacking the gas to form new stars. Both galaxies are around 120 million light years away, and NGC 680 is being distorted by its neighbour's gravity.

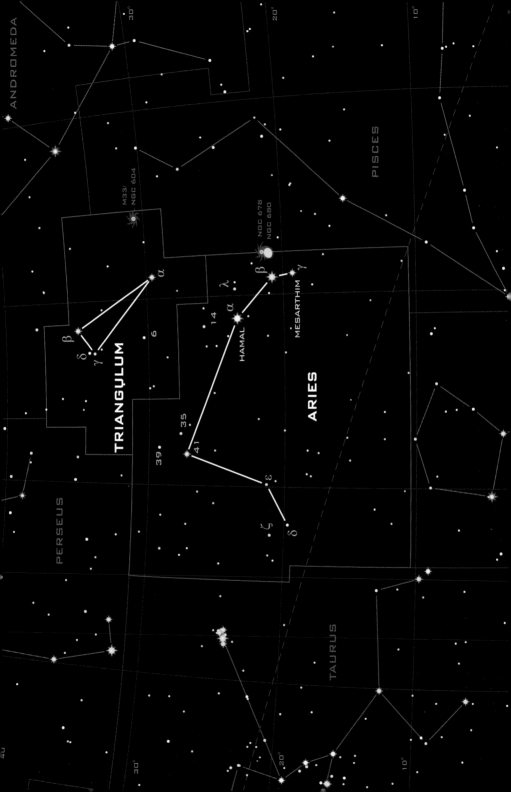

Triangulum Triangulum Galaxy M33

Triangulum Galaxy
Messier 33

Triangulum's greatest showpiece lies close to its western edge. Messier 33 is the closest major galaxy after the Andromeda spiral, Messier 31, and unlike M31 it presents itself face-on to Earth so that its entire spiral structure is displayed. Unfortunately, this means that the galaxy's light is spread out across an area of sky larger than the Full Moon, making it surprisingly hard to see. It is best located in very dark skies using binoculars or a telescope on very low power to spot a contrasting patch of light against the dark sky. Triangulum's disappointingly faint appearance is not entirely due to our viewing angle – it really is a far less impressive galaxy than either M31 or the Milky Way. With a much fainter and looser spiral structure than its near neighbours in the Local Group, M33 is classed as a 'flocculent' spiral – only telescopes of moderate size will be able to pick out any detail within it.

R.A. 01h 34m, DEC. +30º39'
MAGNITUDE 5.7
DISTANCE 2.7 million light years

Infrared View

This view of Triangulum from NASA's Spitzer Space Telescope reveals a range of infrared (heat) radiation from objects too faint to shine in visible light. Red indicates the longest-wavelength, coolest radiations, while green and blue indicate successively shorter wavelengths and warmer material. Stars (including some foreground stars within our own galaxy) appear blue, while colder interstellar dust within the Triangulum Galaxy glows red. While the central blue haze coincides with the extent of the visible galaxy, tendrils of cooler red gas and dust extend well beyond M33's visible limits. This invisible material seems to have migrated outwards from the galaxy's inner regions, but astronomers are still puzzling over exactly how this could have happened.

Fountain of Starbirth
NGC 604

While the Triangulum Galaxy lacks any regimented structure in its star-forming regions, it is still home to some enormous starbirth nebulae. These include NGC 604, one of the largest nebulae in the Local Group. Shown here in a Hubble Space Telescope image, it has an estimated diameter of 1,500 light years. Its diameter is around 40 times that of Orion's Great Nebula, and it is thought to be more than 6,000 times as luminous as that famous starbirth region.

R.A. 01h 34m, **DEC.** +30º47'
MAGNITUDE 14.0
DISTANCE 2.7 million light years

Taurus

AT A GLANCE

NAME Taurus
MEANING The bull
ABBREVIATION Tau
GENITIVE Tauri
R.A. 04h 42m
DEC. +14° 53'
AREA 797 (17)
BRIGHTEST STAR Aldebaran (α)

One of the most distinctive constellations in the sky between Aries to the west and Gemini to the east, Taurus represents the figure of a celestial bull charging towards the hunter Orion. Rich in interesting stars and deep-sky objects, it is also a zodiac constellation that plays host to the Moon and planets.

Lying a little way to the north of the celestial equator, Taurus is visible from nearly all inhabited parts of the world, and is a familiar fixture of northern hemisphere skies on autumn and winter nights. It is one of the few constellations that actually resembles the figure after which it is named – the forequarters of a charging bull – and has been recognized by stargazers since prehistoric times.

The distinctive V-shape of the Hyades star cluster represents the face of the bull, with brilliant red Aldebaran, a foreground star, marking its eye. Meanwhile, two bright stars (one of which, Beta Tauri or Alnath, is shared with neighbouring Auriga) indicate the tips of its horns, fainter chains of stars delineate the beast's forelegs, and the Pleiades star cluster indicates the location of its shoulders.

Face of the Bull
The V-shaped Hyades is the most prominent star cluster in the entire sky – close enough for individual members to be easily distinguished, but far enough away to form a distinct and compact grouping. Evidence from an array of measurements suggests the cluster is around 150 light years from Earth, although the brilliant orange foreground star Aldebaran is at only half this distance. The cluster contains several hundred stars that formed around 600 million years ago.

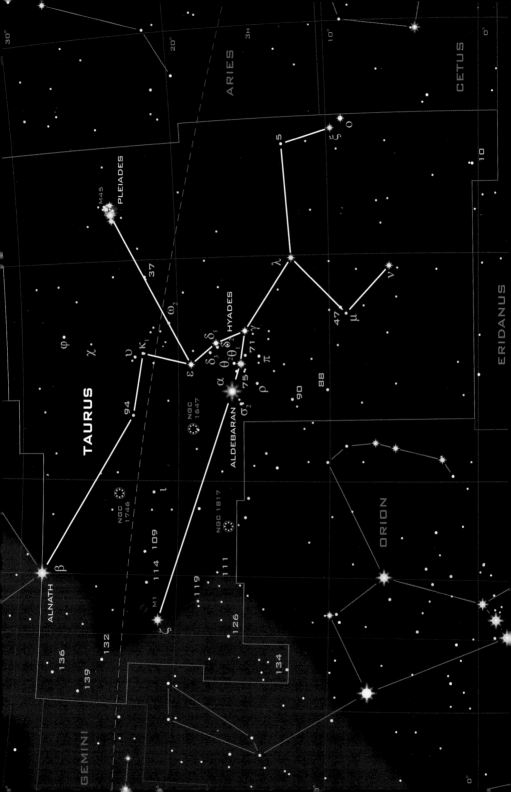

Taurus The Pleiades M45

Seven Sisters

Named for the daughters of Atlas and Pleione in Greek mythology, the Pleiades form an obvious hook-like shape on the shoulder of Taurus. Although clearly visible with the naked eye, the group is not quite as bright as their integrated magnitude of 1.6 would suggest. Observers with average eyesight can usually count six of the 'Seven Sisters', but those with keen eyes can spot seven or even more. Binoculars or a small telescope with low magnification can reveal the cluster in all its glory – it is thought to contain at least 1,000 members within a region around 100 light years across.

R.A. 03h 47m DEC. +24°07'
MAGNITUDE 1.6
DISTANCE 440 light years

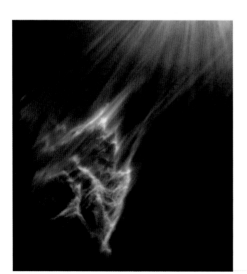

Merope Nebula
NGC 1435

Also known as Tempel's Nebula, NGC 1435 marks the densest concentration of gas and dust in the vicinity of the Pleiades. Shining bluish-white thanks to light scattered from nearby Merope, it forms a patch of light roughly the size of the Full Moon. However, thanks to its faint apparent magnitude of around 13.0, it is only visible through large telescopes or in long photographic exposures.The Hubble Space Telescope image (left) captures the ghostly appearance of the brightest knot within the nebula, a region given its own classification as IC 349.

R.A. 03h 46m, DEC. +23°54'
MAGNITUDE 13.0
DISTANCE 440 light years

Infrared Pleiades

This unusually colourful view of the Pleiades was captured using NASA's infrared Spitzer Space Telescope, and traces density variations in the veil of gas that surrounds the Pleiades Cluster. Red colours highlight the densest regions of the gas cloud, while yellows and greens mark its tenuous outlying regions. This image shows the regions around three of the bright 'sister' stars – Alcyone (centre), Maia (top left) and Merope (top right, embedded in the densest area of nebulosity). Despite appearances, the material around the Pleiades seems not to be linked to their origin – instead it is a separate region of interstellar gas, through which the young cluster just happens to be drifting.

Taurus The Crab Nebula M1

Crab Pulsar
PSR B0531+21

At the heart of the Crab Nebula lies a rapidly spinning neutron star. This is the surviving core of the massive star whose explosion was seen in AD 1054, compressed to the size of a city. Because the collapsed core retains much of the angular momentum and magnetic field of its parent star, it spins at a tremendous rate and channels most of its radiation into two narrow beams aligned with its magnetic poles. As these beams sweep around the sky like a cosmic lighthouse, they are briefly pointed towards Earth, creating a flashing beacon of radio waves known as a pulsar. The Crab Pulsar has now been detected in visible light and X-rays, as shown in this composite image from the Hubble Space Telescope and Chandra X-Ray Observatory.

X-Ray View

An image from NASA's Chandra X-Ray Observatory, a space-based telescope, maps high-energy radiation from the region around the Crab Pulsar. Two jets of X-ray-emitting particles can be seen clearly emerging from above and below the pulsar itself, while a series of concentric waves ripple out into a general X-ray-emitting haze around the collapsed star. The X-rays are a form of synchrotron radiation, emitted by electrons travelling at high speed, and this emission cloud is thought to play a crucial role in transferring energy from the pulsar itself into the nebula. Particles falling onto the rapidly spinning stellar remnant are accelerated to high speeds and ricochet back out to energize the surrounding gas clouds.

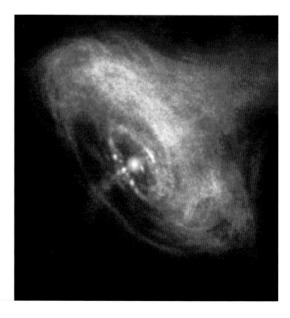

Stellar Remnant

At magnitude 8.4, the Crab Nebula is bright enough to be spotted through binoculars or a small telescope as a fuzzy patch of light northwest of Zeta Tauri (the tip of Taurus's southern horn). Larger instruments or long-exposure images transform it into a shredded web of glowing gas some 11 light years across – the most celebrated supernova remnant in the sky. The Crab was first recorded by British astronomer John Bevis in 1734, and a few decades later French comet hunter Charles Messier listed it as number 1 in his catalogue of objects that might be mistaken for comets. It was not until 1939 that astronomers first associated it with the enormous stellar outburst recorded by stargazers around the world in AD 1054.

R.A. o5h 35m, DEC. +22°01'
MAGNITUDE 8.4
DISTANCE 6,500 light years

Gemini

AT A GLANCE

NAME Gemini
MEANING The twins
ABBREVIATION Gem
GENITIVE Geminorum
R.A. 07h 04m
DEC. +22° 36'
AREA 514 (30)
BRIGHTEST STAR Pollux (β)

The bright stars Castor and Pollux draw attention to this constellation, situated on the zodiac between Cancer and Taurus – small wonder that it has been associated with twins or pairs by cultures across the world. Ancient Chinese astronomers saw these two stars as the balanced principles of yin and yang.

The Romans associated the two stars with the twins Romulus and Remus, founders of Rome itself. However, their common names represent the twin sons of the Spartan Queen Leda, mortal Castor and immortal Pollux, who joined Jason's crew in the quest for the golden fleece.

Fittingly, Castor is a celebrated multiple star, about 52 light years from Earth, with a combined naked-eye magnitude of 1.6. Small telescopes can split it to reveal a pair of blue-white stars of magnitudes 1.9 and 2.9, and reveal a third, more distant red dwarf of magnitude 9.3. Each of these stars is a double in its own right, although the components cannot be separated visually. Despite its designation as Beta Geminorum, Pollux is actually the brighter of the twin stars, shining at magnitude 1.2. In contrast to Castor, it is a stellar singleton – an orange giant around 34 light years away.

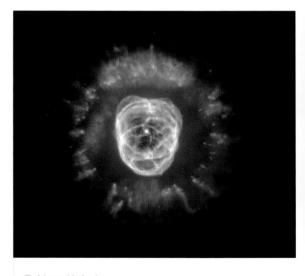

Eskimo Nebula
This complex planetary nebula, also known as NGC 2392, is visible through a small telescope to the southeast of Delta Geminorum. It lies 3,000 light years away and shines at magnitude 10.1. Despite its disc-like appearance, the nebula actually consists of two lobes of material being blown away from the central dying star – one directly towards us, and the other in the opposite direction.

Cancer

AT A GLANCE

NAME Cancer
MEANING The crab
ABBREVIATION Cnc
GENITIVE Cancri
R.A. 08h 39m
DEC. +19° 48'
AREA 506 (31)
BRIGHTEST STAR Al Tarf (β)

The faintest of the zodiac constellations, Cancer was known in medieval times as the 'Dark Sign' on account of its obscurity. It is best found by looking between the brighter stars of Leo and Gemini. Located at the centre of constellation is the Beehive (M44), one of the closest open clusters to Earth.

Despite its faintness, Cancer has been associated with a crab since ancient times. Its origins are untraceable, but the Greeks saw it as a crab that was crushed beneath the foot of Hercules as he fought the serpent Draco. The Egyptians, however, saw the constellation as a sacred scarab beetle.

Despite its designation, Alpha Cancri or Acubens is actually the constellation's fourth-brightest star at magnitude 4.3. A moderate telescope reveals that this white star, 174 light years from Earth, has a faint companion of magnitude 11.9. Zeta Cancri, meanwhile, is a quadruple system – small telescopes show two components of magnitudes 5.1 and 6.2, moderate instruments reveal that the brighter of these is itself an evenly matched double and professional instruments show that the fainter star is itself accompanied by a dim red-dwarf companion.

Beehive Cluster

Cancer's great claim to fame is the star cluster Messier 44, known as the Beehive or Praesepe (the manger). Composed of around 200 stars spread across an area roughly three times the diameter of the Full Moon, the Beehive lies around 580 light years from Earth, and is clearly visible to the naked eye. The great Italian astronomer Galileo Galilei was the first person to resolve its individual stars in the early 1600s, using one of his early, self-built telescopes.

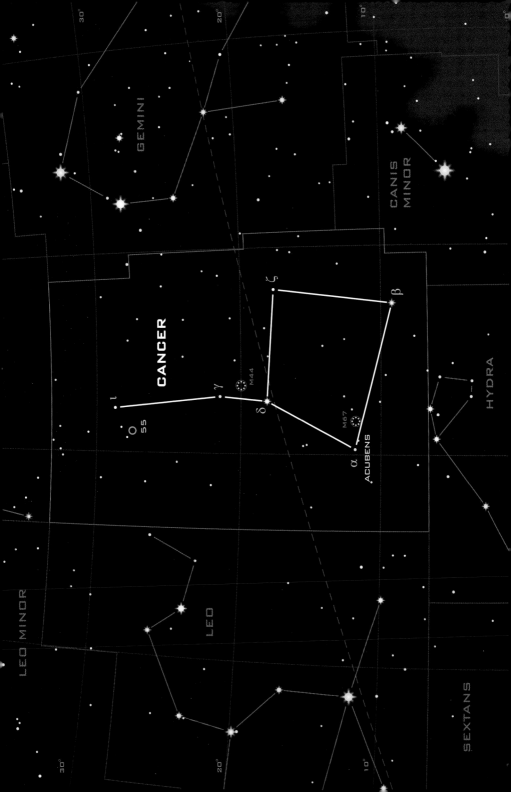

Leo and Leo Minor

AT A GLANCE

NAME Leo/Leo Minor
MEANING The lion/The lesser lion
ABBREVIATION Leo/LMi
GENITIVE Leonis/Leonis Minoris
R.A. 10h 40m/10h 15m
DEC. +13° 08'/+32° 08'
AREA 947 (12)/232 (64)
BRIGHTEST STAR Regulus (α)/
Praecipua (46)

The zodiac constellation of the Lion is one of the most instantly identifiable in the sky. In classical myth it represented the Nemean Lion killed by Hercules. It is a rare case of a constellation that really looks like the object it claims to represent. Sadly, the same cannot be said for its smaller neighbour.

Leo has been recognized as the figure of a lion by almost every culture (though Chinese astronomers saw it as a horse). Since classical times, it has been associated with the Nemean lion fought by the hero Hercules as one of his 12 labours. A curve of stars marking its forequarters is widely known as the Sickle. At its base lies Leo's brightest star, Regulus, whose name means 'little king'. This brilliant white star of magnitude 1.35 lies about 77 light years from Earth. Small telescopes reveal the brighter of two companions that orbit it at a considerable distance. Regulus's location, almost exactly on the ecliptic, allows the Moon and planets to occasionally pass directly in front of it, creating a beautiful event known as an occultation.

In contrast to Leo, faint and shapeless Leo Minor is a late addition to the skies, invented by Polish astronomer Johannes Hevelius in the 17th century.

Hickson 44
Located in the neck of the lion between Gamma and Zeta Leonis, Hickson Compact Group 44 is a small cluster of four galaxies, bound together in a gravitational waltz around 60 million light years from Earth. Its brightest members, the edge-on spiral NGC 3190 and the elliptical NGC 3193 (both in the upper left corner) can be spotted with a small amateur telescope.

Leo Inside View

Barred Spiral
Messier 95

This small but exquisite spiral galaxy is one member of the 'Leo I' group that lies to the south of the lion's body, about halfway along its length. It is a particularly extreme example of a barred spiral with a 'circumnuclear ring' – a circular zone of star formation roughly 2,000 light years across that runs all around the nucleus connecting the ends of the central bar and extending out into the galaxy's spiral arms. Our own Milky Way galaxy, recently revealed as a barred spiral itself, is thought to have a similar circumnuclear ring.

R.A. 10h 44m, DEC. +11º42'
MAGNITUDE 11.4
DISTANCE 38 million light years

Galaxy on Edge
NGC 3628

This unbarred spiral galaxy, seen almost edge-on from Earth, is a member of the Leo Triplet galaxy group along with Messier 65 and M66. While smaller telescopes can show it as an elongated smudge of light, larger instruments reveal a dark lane of dust, associated with the outer edge of the spiral arms, that runs down the length of the galaxy. Although unseen in this image, a long 'tidal tail' extends from one side of the galaxy, reaching more than 300,000 light years into intergalactic space.

R.A. 11h 20m, DEC. +13º35'
MAGNITUDE 9.4
DISTANCE 35 million light years

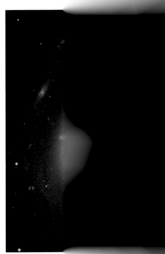

Imperfect Spiral
Messier 96

The largest and brightest galaxy in the Leo I group, M96 is still beyond the reach of binoculars, but just about detectable through a small telescope. It forms a stark contrast to the perfection of its neighbour M95, with warped and faint spiral arms, a nucleus displaced from the centre and an asymmetric distribution of gas and dust. An enticing feature, revealed in this photograph from the European Southern Observatory's Very Large Telescope, is that it lies in front of a more distant galaxy cluster, including a beautiful edge-on spiral at upper left.

R.A. 10h 47m, DEC. +11º49'
MAGNITUDE 10.1
DISTANCE 32 million light years

Warped Beauty
Messier 66

Leo's most spectacular galaxy is Messier 66, the largest and brightest member of the so-called 'Leo Triplet' group that nestles below Theta Leonis in the lion's hindlimb. With a diameter of around 95,000 light years, it is roughly the same size as the Milky Way, but close encounters with its galactic neighbours in the past have warped its disc and spiral arms. They have also concentrated much of the galaxy's mass around its nucleus, while apparently pulling the nucleus away from the geometric centre of the spiral.

R.A. 11h 20m, DEC. +13º00'
MAGNITUDE 8.9
DISTANCE 35 million light years

Coma Berenices

AT A GLANCE

NAME Coma Berenices
MEANING Berenice's hair
ABBREVIATION Com
GENITIVE Comae Berenices
R.A. 12h 47m
DEC. +23° 18'
AREA 386 (42)
BRIGHTEST STAR Beta (β)

Though faint and unremarkable to the naked eye, this constellation reveals its beauty in dark skies or through binoculars. It is home to one of the closest star clusters to Earth, and distant galaxies that can be seen with minimal dust obscuration because it does not lie in the direction of the galactic plane.

Coma, as it is usually known for short, is a rare example of a constellation named to honour a historical, rather than mythological, figure. Queen Berenice was the wife of Ptolemy III, ruler of Egypt in the third century BC, who made an oath to the goddess Aphrodite promising to cut off her luxurious tresses in return for her husband's safety on a distant military campaign. When Ptolemy duly returned home, this starry patch of sky was renamed in her honour.

Alpha Comae Berenices, or Diadem, is a yellow-white star of magnitude 4.3, about 47 light years from Earth and slightly fainter than Beta. Large telescopes show that it is a binary system, with evenly matched stars locked in a 26-year orbit around each other. Aside from the nearby star cluster Melotte 111, Coma lies in a relatively empty region of the sky, allowing views into intergalactic space towards the famous Coma galaxy cluster.

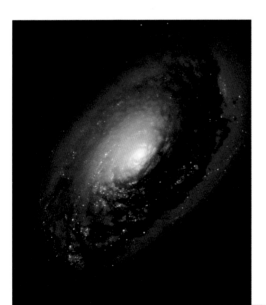

Black Eye Galaxy

At 24 million light years from Earth, Messier 64 is a relatively nearby galaxy that is not assocated with either the Virgo or Coma clusters. Its complex structure includes a thick dust lane that gives it an unmistakable appearance, and inner and outer regions that rotate in opposite directions. These features are thought to be signs that M64 absorbed a smaller satellite galaxy roughly a billion years ago, triggering a wave of star formation.

Coma Berenices Inside View

Unexpected Spiral
NGC 4911

This beautiful Hubble Space Telescope image shows a large, face-on spiral galaxy near the centre of the Coma Cluster about 320 million light years from Earth. NGC 4911 is something of a mystery since it is clearly a site of considerable star-forming activity, revealed by the intense pink nebulae that mark out its spiral arms. Such activity is unusual at the centre of a major galaxy cluster, where intergalactic collisions and close encounters tend to strip galaxies of their star-forming materials.

R.A. 13h 00m, DEC. +27⁰47'
MAGNITUDE 12.8
DISTANCE 320 million light years

Needle Galaxy
NGC 4565

This immaculate edge-on galaxy lies embedded in the midst of the Coma star cluster Melotte 111, but seems to be independent of both the Virgo and Coma galaxy clusters. Discovered by William Herschel in 1785, it is thought to be a barred spiral, although the bar is hidden from our direct view. At magnitude 10.4, it can be located with a small telescope, while a medium-sized instrument will show the dark dust lane running along its length and the central bulge of the nucleus.

R.A. 12h 36m, DEC. +25⁰59'
MAGNITUDE 10.4
DISTANCE 40 million light years

Ring of Starbirth
NGC 4314

This unusual barred spiral galaxy has recently undergone a remarkable burst of star formation. This can be traced through the bright ring around the core, shown in blue and purple in this Hubble Space Telescope image. The ring, 1,000 light years across, is highlighted by the intense blue of young star clusters, and the purple glow of star-forming hydrogen gas.

R.A. 12h 22m, DEC. +29⁰53'
MAGNITUDE 11.4
DISTANCE 40 million light years

Coma Cluster

While southern Coma Berenices plays host to many galaxies that properly belong in the Virgo Cluster, the northern part of the constellation is also home to Coma's own galaxy cluster. The Coma Cluster is far more distant, at around 320 million light years from Earth. Its brightest galaxies are the NGC 4874 and 4889, both giant ellipticals formed from the merger and absorption of numerous smaller galaxies in the cluster's packed central regions. This false-colour view combines a visible-light image in blue with infrared data in red and green.

R.A. 13h 00m, DEC. +27°58'
MAGNITUDE 11.4
DISTANCE 320 million light years

Virgo

AT A GLANCE

NAME Virgo
MEANING The maiden
ABBREVIATION Vir
GENITIVE Virginis
R.A. 13h 24m
DEC. −04° 10'
AREA 1,294 (2)
BRIGHTEST STAR Spica (α)

The second-largest constellation in the sky, Virgo is highlighted by the presence of the brilliant star Spica (Alpha Virginis). It is the 15th brightest star in the sky, with a magnitude of 1. However, Virgo's greatest treasures are telescopic, for it plays host to a wealth of relatively nearby galaxies.

Virgo has been associated with a fertility or harvest goddess since ancient times. The first Babylonian astronomers already associated these stars with the image of a maiden holding an ear of grain or corn (represented by Spica). The Greeks saw her as Persephone, daughter of the harvest goddess Ceres, but also interpreted her as Dike, the goddess of justice holding the scales (Libra) in her raised hand.

Spica lies very close to the ecliptic and is a multiple star system. Its brightest elements, shining at a combined magnitude 1.0, are twin, hot blue-white stars 260 light years away, orbiting each other in just four days and far too tightly bound to be separated by even the largest telescopes. The twin yellow-white stars of Gamma Virginis, or Porrima, on the other hand, can be separated even with small telescopes.

Galaxy Cluster
Northern Virgo is home to the closest major galaxy cluster to Earth, centred around 54 million light years away. Galaxies are gregarious by nature, and the Virgo cluster contains several dozen major spirals and ellipticals, centred around the giant elliptical Messier 87, which is shown here.

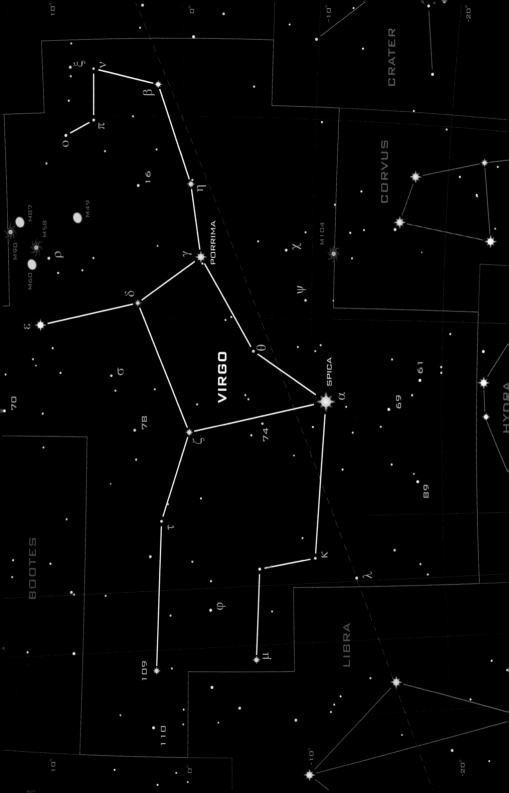

Virgo The Virgo Cluster

Messier 87

The Virgo Cluster's central giant elliptical galaxy is the closest of its kind to Earth. Through small telescopes it looks somewhat similar to a globular cluster, but in reality it is one of the largest known galaxies. A million million stars jostle in a core region roughly 160,000 light years in diameter, while many more orbit in a halo that stretches out to 250,000 light years from the centre. Radio signals and a visible jet of material shooting from the core show that M87 has an active nucleus – a supermassive black hole feeding on material from its surroundings. Hubble Space Telescope images have revealed the dusty skeleton of a cannibalized galaxy embedded within M87, and this material is probably responsible for the ongoing activity.

R.A. 12h 31m, DEC. +12º24'
MAGNITUDE 8.6
DISTANCE 60 million light years

Messier 58

This barred spiral galaxy, 68 million light years from Earth, lies on the far side of the Virgo Cluster, but is one of its brightest galaxies. Small telescopes can locate its bright nucleus, but may have difficulty distinguishing the galaxy from nearby ellipticals – larger instruments are needed to pick up the light of the spiral arms. This infrared image from the Spitzer Space Telescope distinguishes between relatively sedate, mature stars spread throughout the galaxy's central core, bar and disc (shown in blue), and gas and dust in star-forming regions concentrated in the spiral arms (shown in red).

R.A. 12h 38m, DEC. +11º49'
MAGNITUDE 9.7
DISTANCE 68 million light years

Central Regions

This mosaic view covers a broad span of the Virgo Cluster with an area equivalent to 12 Full Moons, revealing two clear concentrations of galaxies within. At lower left, dozens of galaxies cluster around the giant elliptical M87, while at upper right, they gather around two smaller ellipticals, M86 and M84. A third concentration, not shown in this image, is focused on another large elliptical called M49. The three groups are in the process of merging together into a single giant cluster – a process that all clusters eventually pass through and which can help astronomers estimate their age.

R.A. 12h 30m, DEC. +08º00'
MAGNITUDE 9.4
DISTANCE 68 million light years

Virgo The Sombrero Galaxy M104

Silhouetted Ring

The Sombrero's outer dust lane is silhouetted starkly against the galaxy's brilliant core in this stunning Hubble Space Telescope image. The ring has an estimated diameter of around 50,000 light years, making it roughly half the size of our own galaxy's main disc. Nevertheless, at a distance of 28 million light years, the Sombrero appears as one of the brightest galaxies in Earth's skies. It is visible through a small telescope or good binoculars, but larger instruments or long-exposure photographs are needed to identify features such as the central bulge or the dust lane.

Embedded Disc

This spectacular galaxy, unrelated to the Virgo Cluster, lies near the constellation's southern edge around 28 million light years from Earth. It appears to be an elliptical ball of stars with a bright disc within it, as highlighted by this composite of Spitzer Space Telescope infrared data and Hubble Space Telescope images. The ring contains a mix of cool dust (highlighted in red), but also the majority of cold atomic and molecular hydrogen gas – the raw materials that fuel ongoing star formation. In contrast, the galaxy's central regions are lacking in these materials and therefore lack the signs of ongoing star formation.

R.A. 12h 40m, DEC. -11º37'
MAGNITUDE 9.0
DISTANCE 28 million light years

X-Ray View

This composite adds high-energy X-ray data from NASA's Chandra X-Ray Observatory to a visible-light view from the Hubble Space Telescope. X-ray emissions are associated with violent processes and extreme objects – typically the environments around black holes. The Chandra observations show a number of X-ray sources scattered through the Sombrero's elliptical halo, and a blaze of radiation from the region around the nucleus, where a supermassive black hole is thought to lurk. This active nucleus is subdued compared to more violent objects such as quasars and Seyfert galaxies, so the black hole is probably feeding only slowly on gas that drifts within its grasp.

Libra

AT A GLANCE

NAME Libra
MEANING The scales
ABBREVIATION Lib
GENITIVE Librae
R.A. 15h 12m
DEC. −15° 14'
AREA 538 (29)
BRIGHTEST STAR Zubeneschamali (β)

This faint constellation lies midway between the bright stars Spica in Virgo and Antares in Scorpius. Its brightest star, Zubeneschamali, only has a magnitude of 2.6. As the only sign of the zodiac not to represent a living creature, it has often been associated with its brighter neighbours.

In classical times, Libra did not even exist – it was seen as Chelae Scorpionis, the extended claws of neighbouring Scorpio. Libra was established as an independent constellation around the first century AD, by which time it had switched its allegiance to Virgo. It is now often seen as the scales of justice, upheld by Virgo in her guise of the goddess Iustitia.

Zubenelgenubi (Alpha Librae) is a wide double star, easily separated through binoculars into individual components of magnitudes 2.2 and 5.2. These two stars are 77 light years from Earth, orbiting each other in around 200,000 years. The bright primary star is a binary in its own right, although it is inseparable through even the largest telescopes. Zubeneschamali (Beta Librae) meanwhile, is one of only a very few stars in the sky to display a noticeably greenish colour.

Gliese 581

This red dwarf to the northeast of Beta Librae lies just 22 light years from Earth and is home to one of the closest known planetary systems. A Neptune-sized planet was discovered here in 2005, and since then at least two, and possibly as many as five more, have been identified. These include Gliese 581c, one of the most Earth-like exoplanets known, and Gliese 581d, which orbits in its solar system's 'habitable zone' where liquid water can persist on its surface.

Serpens

AT A GLANCE

NAME Serpens
MEANING The serpent
ABBREVIATION Ser
GENITIVE Serpentis
R.A. 16h 57m
DEC. +06° 07'
AREA 637 (23)
BRIGHTEST STAR Unukalhai (α)

Uniquely in the sky, this constellation is split into two distinct parts – Serpens Caput (the snake's head) and Serpens Cauda (the snake's tail) – on either side of the larger constellation Ophiuchus (the Serpent Bearer). Its brightest star is variously identified as either the neck or the heart of the serpent.

This region of the sky has been interpreted as a figure wrestling a dragon since the earliest times. Ancient Greek astronomers associated it with a more benign snake wielded by Asclepius, god of medicine. Serpens Cauda is home to a rich region of the Milky Way including the spectacular Eagle Nebula and its associated star cluster Messier 16, home to the celebrated 'Pillars of Creation'. Serpens Caput, meanwhile, looks towards a region of space above the plane of the Milky Way, and is home to the bright globular cluster M5.

Alpha Serpentis, or Unukalhai, (the 'serpent's neck'), is an orange giant 73 light years from Earth, shining at magnitude 2.6. Delta Serpentis is a multiple-star system consisting of a brighter pair (magnitudes 4.2 and 5.2) 210 light years away, orbited by a much fainter pairing of red dwarfs visible only through larger telescopes.

Serpens South

Dense clouds of dust obscure parts of the Milky Way in southern Serpens Cauda, at the base of the long Cygnus Rift (see page 53). In 2007, scientists used the infrared vision of NASA's Spitzer Space Telescope to peer through the dust and discover a young star cloud beyond. The Serpens South Cluster consists of around 50 stars crammed into a region roughly five light years across, some 850 light years from Earth, including many that are still in the process of formation.

Serpens Inside View

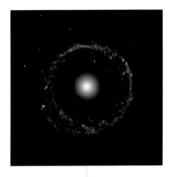

Hoag's Object

When US astronomer Art Hoag discovered this beautiful wheel in space in northwestern Serpens Caput in 1950, he was unsure whether it was a nearby planetary nebula or an unusual distant galaxy. Later observations confirmed that Hoag's Object is in fact a rare 'ring galaxy' 800 million light years from Earth. Other ring galaxies have since been identified, but they usually have features that can be traced back to a collision with a smaller galaxy – the shock wave triggers a ring of star formation. Hoag's Object, however, seems to have formed without such interactions – astronomers are still unsure of its origin.

R.A. 15h 17m, DEC. +21^{0}35'
MAGNITUDE 0.3–1.2 (var)
DISTANCE 800 light years

Messier 5

Just visible to the naked eye in dark skies, M5 is the brightest of several globular clusters scattered across Serpens and neighbouring Ophiuchus. It appears as an impressive fuzzy 'star' through binoculars and a small telescope will begin to resolve individual stars in its outer reaches. The cluster is about 165 light years across and contains several hundred thousand stars – perhaps as many as half a million. It is also one of the oldest known globular clusters at 13 billion years old, and is particularly rich in variable stars of the 'RR Lyrae' type, which have been used to accurately measure its distance.

R.A. 15h 19m, DEC. +02^{0}05'
MAGNITUDE 5.6
DISTANCE 24,500 light years

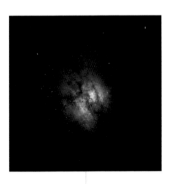

Arp 220

This unusual object in Serpens is known by its catalogue entry in the Atlas of Peculiar Galaxies published by US astronomer Halton Arp in 1966. Arp identified 338 galaxies whose structure did not conform to normal classifications such as spiral, elliptical and irregular. In the ensuing decades, improved observations have shown that many of his peculiar objects are in fact pairs of colliding galaxies, and satellite observatories have shown other unusual features. Arp 220, for example, emits extremely large amounts of infrared (heat) radiation, thought to be generated by massive bursts of star formation as the galaxies collide and merge.

R.A. 15h 34m, DEC. +23^{0}30'
MAGNITUDE 13.9
DISTANCE 250 million light years

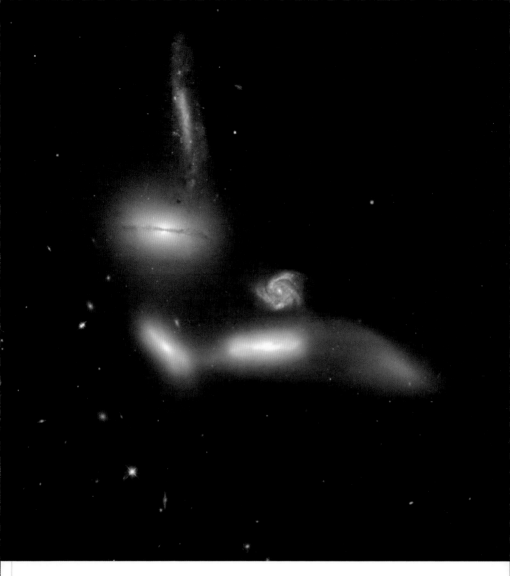

Seyfert's Sextet
NGC 6027

Discovered by US astronomer Carl Seyfert in the late 1940s, this compact galaxy group in Serpens Caput was at first thought to contain six galaxies. However, later studies showed that the small face-on spiral is a far more distant background object, and the diffuse, fuzzy object at lower right is a 'tidal tail' of material pulled away from one of the other interacting galaxies. Nevertheless, the remnant cluster is still impressive, with four galaxies (three edge-on spirals and one elliptical in the centre with a prominent dust lane) jammed together in a region of space just 100,000 light years across.

R.A. 15h 59m, **DEC.** +20º45'
MAGNITUDE 14.7
DISTANCE 190 million light years

Serpens The Eagle Nebula M16

Eagle Nebula
Messier 16

The star cluster M16 hovers just on the edge of naked-eye visibility at magnitude 6.4, but is easily identified with binoculars or a small telescope. The cluster is very young, at just 5 million years old, and is still embedded in the spectacular Eagle Nebula, a site where star formation is still continuing. While Swiss astronomer Philippe Loys de Chéseaux discovered the cluster in 1745, it was Charles Messier who discovered the surrounding nebula some 19 years later. Despite its impressive appearance in photographs, the Eagle will only reveal its secrets in detail to the largest amateur telescopes.

R.A. 18h 19m, DEC. -13º47'
MAGNITUDE 6.4
DISTANCE 7,000 light years

Pillars of Creation

In 1995, the Hubble Space Telescope captured this iconic image of dense star-forming columns in the heart of the Eagle Nebula. It provided astronomers with their first detailed view of the processes involved in star formation, and was soon dubbed the 'Pillars of Creation' – a name enthusiastically adopted by the media. The individual pillars in the image are up to four light years long – they mark regions whose density is high enough to resist the pressure of radiation from the bright stars in M16 forcing back the walls of gas within the nebula. Inside each of them, dozens of individual knots of gas are growing denser still, developing their own gravity so that they can pull more material from their surroundings and snowball into newborn stars.

The Spire

In 2005, Hubble scientists returned to the Eagle Nebula to study another of its features – a longer and more tenuous column of gas nicknamed the Spire. The spire is being rapidly eroded by radiation from nearby newborn stars, triggering emission of light that surrounds it with an eerie glow. Dense knots of material inside the hydrogen-rich column of gas may still develop to form stars, but as the raw materials are stripped away from around them their growth will be stunted. The first generations of stars to form in any nebula have a natural advantage when it comes to gaining material, and also pull the ladder up behind them, depriving later stellar generations of available material.

Ophiuchus

AT A GLANCE

NAME Ophiuchus
MEANING The serpent bearer
ABBREVIATION Oph
GENITIVE Ophiuchi
R.A. 17h 24m
DEC. −07° 55'
AREA 948 (11)
BRIGHTEST STAR Rasalhague (α)

This large equatorial constellation appears rather empty of bright stars, and is best found by looking between the bright stars Altair (in Aquila) and Spica (in Virgo), and to the north of brilliant Antares in Scorpius. It contains Barnard's Star, the second nearest star system to Earth at 5.9 light years.

Ophiuchus is intimately associated with the two-part constellation of Serpens, whose head and tail lie to either side of it. Although the constellation is usually depicted as a giant wrestling a snake, the figure has been seen since Roman times as Asclepius, the god of healing, who traditionally carried a staff with a serpent twined around it. The constellation is sometimes known as the 'thirteenth sign of the zodiac' – the effects of precession (*see* page 11) mean that the ecliptic now passes through it and the Moon and planets can often be seen here.

The constellation's brightest star, Rasalhague, is a slightly variable white star of magnitude 2.1. Lying around 47 light years from Earth, it forms a binary system with an unseen companion. Rho Ophiuchi, meanwhile, is a beautiful quadruple star system wth a combined magnitude of 4.6, still embedded in the nebulosity from which it formed.

Messier 9

Ophiuchus contains a wealth of globular clusters, of which Messier 9 is probably the most impressive. Discovered in 1764, it lies close to the centre of our galaxy, 25,800 light years from Earth, and shines at magnitude 8.4. As a result, it is only just visible through binoculars, but makes a fine sight through any telescope.

Ophiuchus

Lobster Galaxy
NGC 6240

Photographed by the Hubble Space Telescope in 2008, this unusual object is the disrupted result of a galactic collision and merger that began around 30 million years ago and may not be complete for another 100 million years. X-ray images have revealed the presence of two distinct nuclei – giant black holes some 3,000 light years apart that are ultimately destined for their own collision and merger. The galaxy pumps out far more infrared (heat) radiation than would normally be expected, perhaps because of the heating effect of accelerated star formation or due to the influence of its active black holes.

R.A. 16h 52m, DEC. +02º24'
MAGNITUDE 12.8
DISTANCE 400 million light years

Barnard's Star
V2500 Ophiuchi

Barnard's Star is an apparently insignificant red dwarf raised to fame by its proximity to Earth (at just 5.9 light years away) and its speed across the sky. It is the star with the highest 'proper motion' – the star's own movement through space, combined with our Solar System's drift in the opposite direction, ensures that it moves half a degree (a Full Moon width) across the sky every 180 years. The finder chart (right) shows its location within Ophiuchus and its general direction of motion. Despite its proximity, Barnard's Star was only discovered by the US astronomer E.E. Barnard in 1916.

R.A. 17H 58M, DEC. +04º42'
MAGNITUDE 9.5
DISTANCE 5.9 light years

Rho (ρ) Ophiuchi

Although unimpressive to the naked eye, a small telescope or even binoculars will transform Rho Ophiuchi into a beautiful multiple system with a central star of magnitude 5.0, and companions of magnitudes 5.9, 6.7 and 7.3. All the components are hot blue-white stars, just a few million years old and still embedded in remnants of the gas from which they formed. Indeed, long-exposure photographs show that the entire region is filled with beautiful glowing gas clouds and silhouetted dust lanes. In this image, Rho Ophiuchi is surrounded by blue nebulosity at the top of the picture, while the orange glow from nearby Antares intrudes at lower left.

R.A. 16h 26m DEC. -23º27'
MAGNITUDE 4.6
DISTANCE 395 light years

Aquila and Scutum

AT A GLANCE

NAME Aquila/Scutum
MEANING The eagle/The shield
ABBREVIATION Aql/Sct
GENITIVE Aquilae/Scuti
R.A. 19h 40m/18h 40m
DEC. +03° 25'/−09° 53'
AREA 652 (22)/109 (84)
BRIGHTEST STAR Altair (α)/Alpha (α)

These two equatorial constellations are easily located thanks to the bright star Altair. For northern observers, Altair forms the base of the 'summer triangle' in association with the nearby bright stars Deneb and Vega. Southern stargazers can find it by looking to the north of the constellation Sagittarius.

Aquila has been seen as an eagle since at least the fourth century BC – it was often associated with the bird that held Zeus's thunderbolts, or with Zeus himself, who transformed into an eagle to abduct Ganymede to serve as a cupbearer. Scutum, meanwhile, was invented by Johannes Hevelius in 1684, and represents the shield of Polish King John III Sobieski.

Altair, or Alpha Aquilae, is the twelfth-brightest star in the sky at magnitude 0.8. Lying just 17 light years from Earth, it is one of the few stars to have had its surface imaged directly. This revealed a bulging equator caused by the young white star's rapid spinning motion. Magnitude-3.8 Alpha Scuti, meanwhile, is an orange giant 174 light years away, seen against some of the brightest Milky Way star clouds. Scutum also contains Messier 11, also known as the 'Wild Duck', one of the Milky Way's finest star clusters.

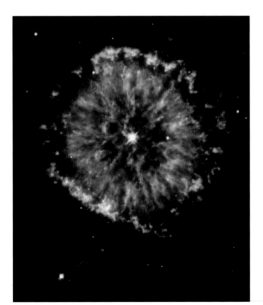

NGC 6751

This attractive but challenging planetary nebula lies close to Lambda Aquilae in the constellation's southwest corner. Shining at magnitude 11.9, it lies some 6,500 light years from Earth and has an estimated diameter of about 0.8 light years. The nebula's complex structure is a result of collisions between fast-expanding gas from the hot central star and cooler material expelled thousands of years ago.

Delphinus and Equuleus

AT A GLANCE

NAME Delphinus/Equuleus
MEANING The dolphin/The little horse
ABBREVIATION Del/Equ
GENITIVE Delphini/Equulei
R.A. 20h 42m/21h 11m
DEC. +11° 40'/+07° 45'
AREA 189 (69)/72 (87)
BRIGHTEST STAR Rotanev (β)/
Kitalpha (α)

Two compact constellations – the diamond-like form of a leaping dolphin and the small skewed rectangle of a horse's head – lie to the southwest of the easily identifiable great Square of Pegasus. The four brightest stars in Delphinus form a diamond-shaped star pattern called 'Job's Coffin'.

Despite its small size and relative simplicity, the stars of Delphinus do indeed bear a distinct resemblance to a dolphin. Mythologically, this creature was said to be a servant of the god Poseidon, sent to rescue the bard Arion from a shipwreck. The constellation's brightest stars, Beta at magnitude 3.6 and Alpha at magnitude 3.8, are both binaries, though neither can be separated with an amateur telescope. Gamma Delphini, however, is an easily observable binary consisting of a yellow-orange primary at magnitude 4.3, and a fainter, magnitude-5.1 companion that can appear either white, blue or greenish in colour.

Equuleus, meanwhile, bears little resemblance to a horse of any kind, and is usually represented as simply an equine head. Despite its obscurity, the constellation dates back to ancient times and is often associated with Celeris, the foal of the mythical winged horse Pegasus.

Distant Giant
Globular cluster NGC 6934 in southern Delphinus was discovered by German-born British astronomer William Herschel in 1785. With a magnitude of 8.8, and a distance of 50,000 light years, it lies on the limit of binocular visiblity but is a good target for small telescopes.

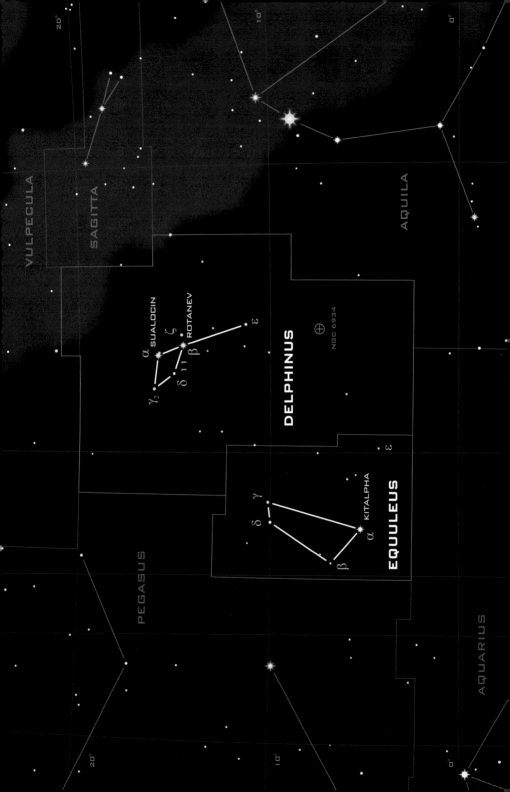

Pegasus

AT A GLANCE

NAME Pegasus
MEANING The winged horse
ABBREVIATION Peg
GENITIVE Pegasi
R.A. 22h 42m
DEC. +19° 28'
AREA 1,121 (7)
BRIGHTEST STAR Enif (ε)

While it bears little resemblance to the figure of a winged horse that it is supposed to represent, this constellation is still unmistakable thanks to the prominent square formed by its brightest stars. In Greek mythology Pegasus was ridden by the Greek heroes Perseus and Bellerophon.

The stars of the 'Square of Pegasus' are usually depicted as the forequarters of the horse, with fainter chains of stars indicating its raised forelimbs and head. The entire depiction is upside-down for northern hemisphere observers. Of the four stars in the square, the northeastern one (Delta Pegasi) is shared with neighbouring Andromeda (see page 58). Alpha, or Markab, is a white star of magnitude 2.5, 140 light years from Earth, while Beta (Scheat) is a red giant 200 light years away, varying between magnitudes 2.3 and 2.7 in an unpredictable cycle. Gamma, or Algenib, in contrast, is a hot blue star about 330 light years away, shining at magnitude 2.8.

The constellation's brightest star, however, is Epsilon Pegasi or Enif. This orange supergiant varies around around magnitude 2.4, and marks the horse's nose. It is 690 light years from Earth, with a blue companion easily seen through a small telescope.

Stephan's Quintet

When French astronomer Edouard Stephan discovered this cluster of galaxies in 1877 (now catalogued as Hickson Compact Group 92), he concluded that it contained five members. But more recent measurements have confirmed that the spiral at upper left is a foreground galaxy – just 40 million light years away compared to the 300 million light years of the four true members.

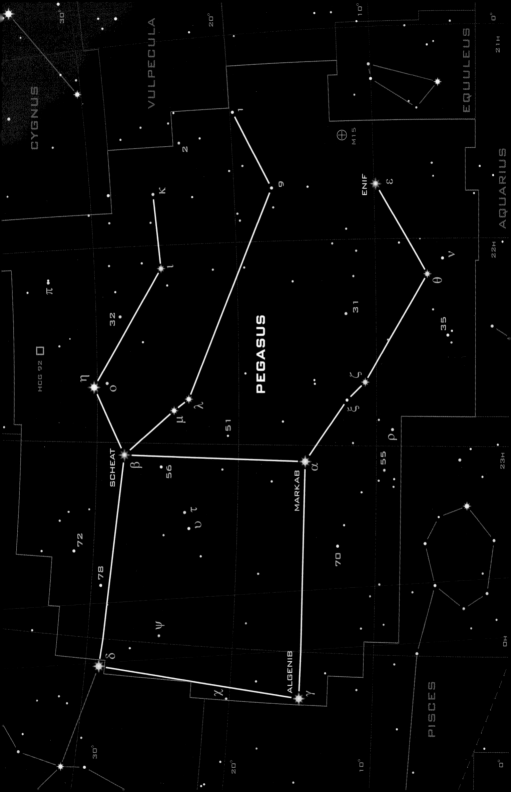

Aquarius

AT A GLANCE

NAME Aquarius
MEANING The water carrier
ABBREVIATION Aqr
GENITIVE Aquarii
R.A. 22h 17m
DEC. −10° 47'
AREA 980 (10)
BRIGHTEST STAR Sadalsuud (β)

This somewhat shapeless constellation is nevertheless one of the oldest. It has been recognized as the figure of a man pouring water from a jug by astronomers since at least the time of ancient Babylon, around 2,000 BC. It was one of the 48 constellations listed by the second century AD astronomer Ptolemy.

Since classical times, Aquarius has been associated with Ganymede, a beautiful youth kidnapped by Zeus (in the form of the eagle Aquila) and taken to Mount Olympus to act as cup bearer to the Greek gods. A Y-shaped group of relatively bright stars around Zeta Aquarii is known as the Water Jug, and the water pours southeast from here towards the bright star Fomalhaut in neighbouring Piscis Austrinus (*see* page 160).

The constellation's brightest stars at magnitude 2.9, Alpha and Beta Aquarii are known as Sadalmelik and Sadalsuud respectively. Both are yellow supergiants 760 and 610 light years from Earth. Beta is less luminous than Alpha, but appears fractionally brighter in Earth's skies thanks to its relative proximity. Zeta Aquarii, meanwhile, is an attractive double star with near-twin white components of magnitudes 4.3 and 4.5, sitting almost exactly on the celestial equator.

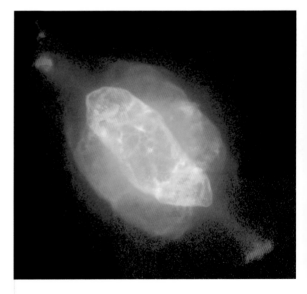

Saturn Nebula
This planetary nebula, catalogued as NGC 7009, lies around 5,200 light years from Earth in western Aquarius, and takes its name from its obvious resemblance, seen through a small telescope, to the ringed planet Saturn. The compact nature of this magnitude-8.0 nebula makes it relatively easy to observe.

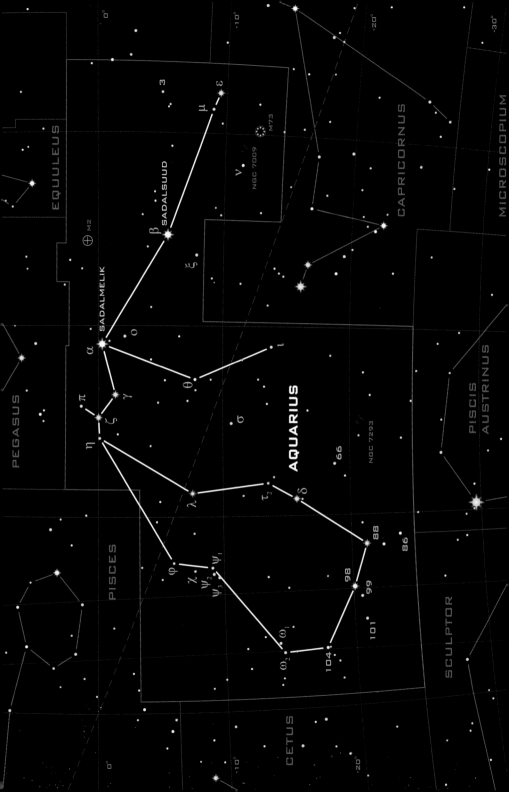

Aquarius The Helix Nebula NGC 7293

Spectacular Vista

This striking portrait of NGC 7293 comes from the European Southern Observatory's VISTA infrared telescope in Chile. While ground-based infrared observations cannot detect the low temperature materials seen by satellites such as Spitzer, they can still offer new perspectives – in this case by 'tuning in' to the radiation emitted by cool molecular gases in and around the nebula. These observations not only reveal previously unsuspected fine structures in the bright part of the nebula, but also show that gas shed from the central star extends far beyond its visible limits, out to distances of around four light years.

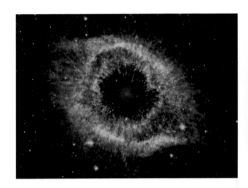

A Different Angle

Astronomers constructed this three-dimensional model of the Helix's structure with the aid of optical and radio studies. By measuring the direction and speed with which different parts of the nebula are moving, they discovered that previous models of a 'bagel-shaped' nebula were wrong. Instead, the Helix appears to be composed of two intersecting discs, tilted at a sharp angle so they are more or less perpendicular to each other. The smaller disc is responsible for the main ring structure we see from Earth, while the larger disc is responsible for features such as the broad outer loop, illuminated as the entire nebula moves through space and collides with interstellar matter.

Infrared Eye

This infrared view of the Helix Nebula combines images taken in different wavelengths to reveal temperature differences within. Blue and green in the outer regions show comparatively hot gas and help pick out the radial structure created as the star blew away its outer layers near the start of its decline from red giant into white dwarf. The red glow closer to the centre marks relatively cool dust, and was an unexpected discovery from these Spitzer Space Telescope observations. One theory is that it was kicked up by a cloud of comets that orbited far enough out from the star to avoid destruction during its red-giant phase.

R.A. 22h 30m, DEC. -20º48'
MAGNITUDE 7.3
DISTANCE 450 light years

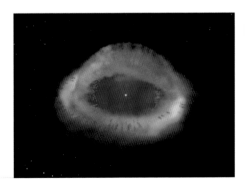

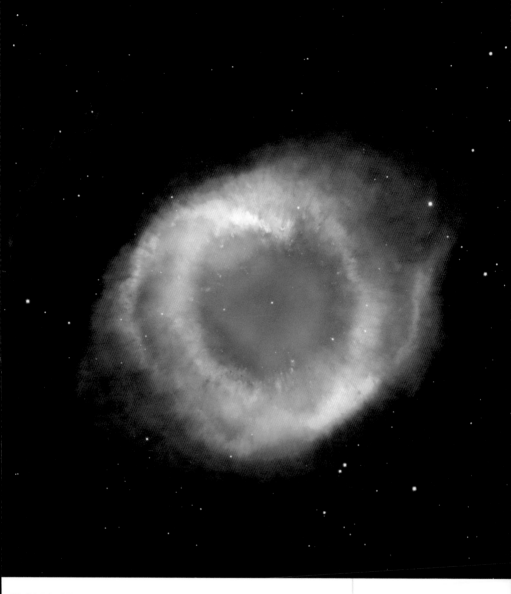

Hubble's View

The Helix is one of the brightest planetary nebulae visible from Earth, but also one of the largest on account of its proximity to Earth. As a result, its light is spread out over an area larger than the Full Moon. For most observers, therefore, the best way of looking for the Helix is with binoculars or a low-magnification telescope, from a site with very dark skies. At best the nebula will look like a hazy disc of light – professional telescopes and long exposures are required to capture the complexities shown in this view from the Hubble Space Telescope.

Cetus

This large constellation, straddling the celestial equator southwest of Taurus, can be challenging to spot on account of its tendency to change appearance depending on the brightness of one of its key stars, Mira. The brightest star, Deneb Kaitos (Arabic for 'tail of the whale'), has a visual magnitude of 2.04.

AT A GLANCE

NAME Cetus
MEANING The sea monster
ABBREVIATION Cet
GENITIVE Ceti
R.A. 01h 40m
DEC. −07° 11'
AREA 1,231 (4)
BRIGHTEST STAR Deneb Kaitos (β)

Strictly speaking, the Latin word *cetus* translates as 'whale', but since ancient times this constellation has been identified as a far more fearsome sea monster called Typhon. This creature plays a role in myths associated with the constellations Pisces and Capricornus, but is best known as the monster sent by Hera, the jealous Queen of the Gods, to ravage the kingdom of Cassiopeia and Cepheus. It is killed by the hero Perseus, who saves the life of Andromeda.

Cetus is home to a variety of interesting objects including the attractive binary star Gamma Ceti, the line-of-sight double Alpha Ceti (Menkar) and the nearby stars Tau Ceti and UV Ceti A and B. However, its most famous star by far is Omicron Ceti, or Mira – a slowly pulsating red giant that varies between around magnitude 3.0 and magnitude 10.0 in a 332-day cycle. At its brightest, Mira joins the head and body of Cetus at the neck, but at its faintest, its disappearance splits Cetus into two distinct star groups.

Speeding Wonder
This ultraviolet image reveals that Mira (whose name means 'wonderful') is racing through space at high speed and leaving a trail of hot gas in its wake. This material has been flung off in each of its long, slow pulsations, and pulled away by the tug of a close binary companion. Theoretical models suggest the star varies between 300 and 400 times the diameter of the Sun in the course of each 332-day cycle.

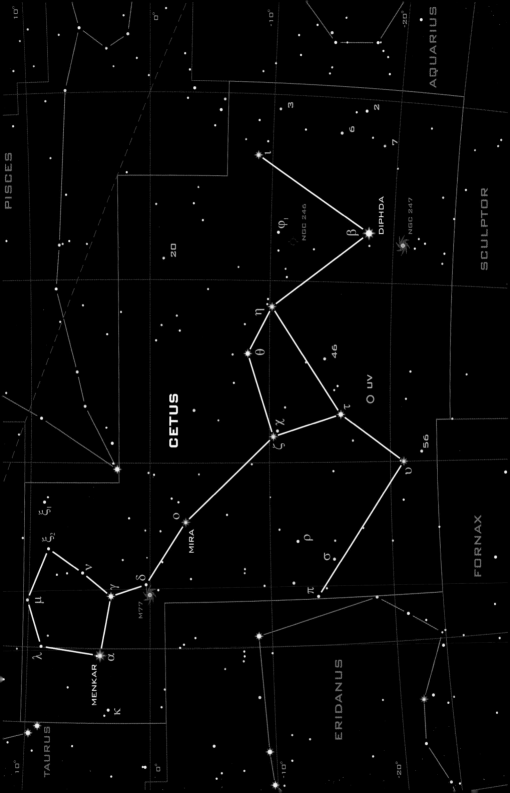

Orion

AT A GLANCE

NAME Orion
MEANING Orion
ABBREVIATION Ori
GENITIVE Orionis
R.A. 05h 35m
DEC. +05° 57'
AREA 594 (26)
BRIGHTEST STAR Rigel (β)

This bright constellation dominates both northern and southern hemisphere skies early in the year. One of the most recognizable star patterns in the night sky, it is studded with celestial marvels ranging from turbulent starbirth nebulae to dying stars near the end of their lives.

Horsehead Nebula
A vast cloud of interstellar material lies behind much of Orion, only becoming visible when excited by the light of nearby stars, or silhouetted against more distant light sources. Perhaps the most famous region of the cloud is the Horsehead Nebula, or Barnard 33, a renowned celestial chesspiece formed by a loop of obscuring dust 3.5 light years wide, visible against the IC 434 nebula.

Orion represents a mighty hunter of Greek and Roman myth, the lover of Artemis, goddess of the hunt. Orion was said to have boasted that he could overcome any creature on Earth, whereupon the Earth brought forward a scorpion that stung him to death. For this reason, the constellations of Orion and Scorpius are today preserved on opposite sides of the sky.

Two of the brightest stars in the sky, Betelgeuse and Rigel, mark the hunter's shoulder and knee, while a chain of three somewhat fainter ones mark his belt. The easternmost of these, magnitude-1.8 Zeta Orionis, or Alnitak, is a multiple star that reveals a companion of magnitude 4.0 with a small telescope, and a third component of magnitude 9.5 through larger instruments. South of Alnitak, runs a chain of stars and nebulosity forming Orion's Sword – one of the most rewarding regions in the entire sky.

Orion Inside View

Betelgeuse
α Orionis

Alpha Orionis is the distinctive red supergiant Betelgeuse. Despite its 'alpha' designation, and its rank as the tenth-brightest star in the sky, it is not the brightest star in Orion – that honour goes to Rigel. Betelgeuse is a fascinating star, however: it is so large that it is unstable and fluctuates in both size (between 300 and 400 times the diameter of the Sun) and brightness (between magnitudes 1.2 and 0.3). The huge size of Betelgeuse also allowed the Hubble Space Telescope to perform a notable first – direct imaging of the surface of a distant star.

R.A. o5h 55m, DEC. +07º24'
MAGNITUDE 0.3–1.2 (var)
DISTANCE 640 light years

LL Orionis

This faint young star originated in the Orion Nebula relatively recently, and is now moving through the nebula's outer reaches, drifting away into space. An impetuous youth, LL Orionis generates a much fiercer stellar wind than more sedate stars like the Sun. As this wind encounters the oncoming pressure of gas in the wider nebula, it generates a curving shock wave called a 'bow shock', analogous to the wave generated in front of a fast-moving ship. Collisions between particles from the nebula and those from LL Orionis's stellar winds release the energy that makes the interface glow.

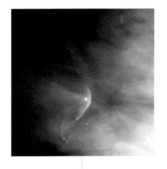

R.A. o5h 35m, DEC. -05º25'
MAGNITUDE 11.5
DISTANCE 1,350 light years

Rigel
β Orionis

Despite its designation, Rigel is the constellation's brightest star. In striking contrast to Betelgeuse, it is a brilliant blue-white colour. Some have speculated that the mislabelling of the two stars suggests Betelgeuse actually outshone Rigel in the historic past. Brilliant blue-white supergiants such as Rigel are some of the most massive stars known – Rigel has an estimated mass of 17 Suns, a surface temperature of 11,000°C (19,800°F) and a luminosity 66,000 times that of our own Sun.

R.A. o5h 14m, DEC. -08º12'
MAGNITUDE 0.1
DISTANCE 860 light years

Flame Nebula
NGC 2024

Lyng just a little way to the north of the famous Horsehead Nebula, the Flame is
another celebrated cloud of interstellar gas and dust. It is illuminated in part by
the radiation that floods out from Alnitak or Zeta Orionis, the easternmost star
of Orion's Belt, and in part by the radiation from new stars being born inside the
nebula. Dark clouds of dust lying in front of the nebula give it the appearance of
a billowing flame in visible light, but this near-infrared view from the European
Southern Observatory's VISTA telescope lifts the veil of dust to reveal a complex
celestial landscape beneath.

R.A. 05h 42m, DEC. -01º51'
MAGNITUDE 2.0
DISTANCE c.900 light years

Orion The Orion Nebula M42

Infant Stars

As one of the closest nearby star factories, the Orion Nebula offers a unique opportunity for astronomers to test their theories and look at the properties of newborn stars. One important feature they have identified is the near-ubiquity of protoplanetary discs or 'proplyds'. These dense clouds of gas and dust persist in orbit around stars even after they have begun to shine, and therefore seem to be an ideal source of raw materials for the later formation of solar systems. Another significant Hubble discovery within the nebula is an abundance of lightweight brown dwarfs – 'failed stars' that never achieved sufficient mass to ignite nuclear fusion reactions, but still glow with heat from their formation.

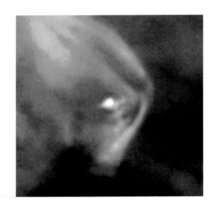

R.A. 05h 35m, **DEC.** -05º27'
MAGNITUDE 4.0
DISTANCE 1,350 light years

Trapezium Cluster

At the heart of the Orion Nebula lies a tight cluster of stars that smaller telescopes show as a quadruple star system. The cluster, universally known as the Trapezium, was first reported by the Italian astronomer Galileo Galilei in his 1610 book *The Starry Messenger*, and while the number of stars in the cluster has steadily increased, the name has stuck. The five dominant stars of the Trapezium, shown in this Hubble Space Telescope image, each have masses between 15 and 30 times that of the Sun. This ensures that they will live fast and die young, but for the moment the ultraviolet radiation they pour out into the surrounding nebulosity does much to illuminate the gas clouds.

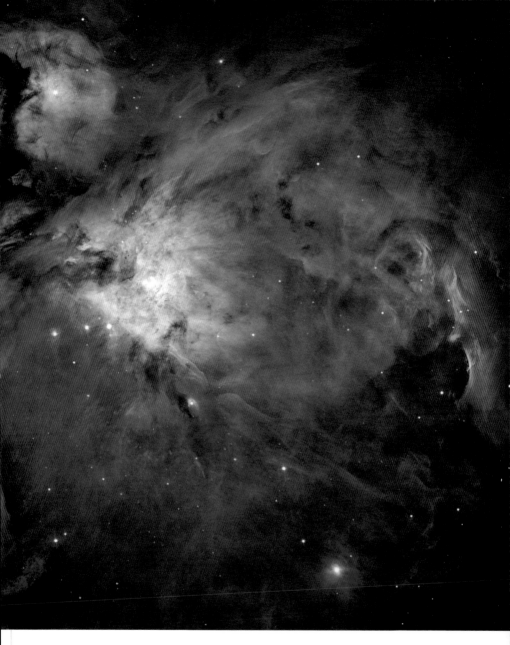

Great Nebula

M42 is easily seen wth the naked eye as a patch of fuzzy light just south of the multiple star system Theta Orionis in Orion's Sword. Many of the nebula's features can already be traced using just binoculars or a small telescope, although the nebula appears greenish rather than pink. Larger telescopes only increase the levels of detail visible in this wonderful flower-like structure. As well as M42, Orion's Sword is home to several other interesting nebulae, though none is as bright and large as the so-called Great Nebula.

Monoceros and Canis Minor

AT A GLANCE

NAME Monoceros/Canis Minor
MEANING The unicorn/The lesser dog
ABBREVIATION Mon/CMi
GENITIVE Monocerotis/Canis Minoris
R.A. 07h 04m/07h 39m
DEC. +00° 17'/+06° 26'
AREA 482 (35)/183 (71)
BRIGHTEST STAR Beta (α)/Procyon (α)

Lying directly to the east of brilliant Orion, the large 'W' of Monoceros provides a much fainter observing prospect. Fortunately, Procyon, the brightest star of Canis Minor, lies directly to its north. Monoceros is Greek for unicorn, the legendary beast with a single horn on its head.

Monoceros stretches across a bright region of the Milky Way, which makes its pattern all the more difficult to distinguish. It is widely thought to have been invented by Dutch theologian Petrus Plancius around 1613, although some early historians claimed it had more ancient Persian or Arabic origins. Despite its obscurity, Monoceros is home to several interesting stars and deep-sky objects – Beta Monocerotis, for instance, is an attractive triple star that is visible through the smallest telescopes.

Canis Minor, in contrast, is a compact region with little to offer apart from the bright star Procyon shining at magnitude 0.3. Despite this, the lesser dog is one of Ptolemy's classical constellations, and has been recognized since the first century BC. Procyon itself is one of the closest bright stars to Earth, just 11.4 light years away and similar in many ways to the even brighter Sirius (*see* page 130).

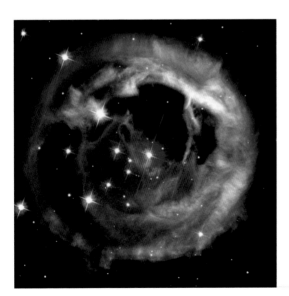

V838 Monocerotis

In January 2002, astronomers spotted an unusual stellar explosion to the southwest of Delta Monocerotis – an object soon designated as V838 Monocerotis. Since then, this strange star has been a subject of intensive study as the 'light echo' of the explosion has reflected off surrounding gas and dust. The origin of the outburst, it seems, was a supergiant star nearing its death in a supernova.

Monoceros Inside View

Hubble's Variable Nebula, NGC 2261

This curious faint nebula lies close to 15 Monocerotis in the north of the constellation and is noted for its strange behaviour, varying in both brightness and apparent shape. It is named in honour of the great US astronomer Edwin Hubble, who was the first person to study it in detail, and it is associated with the unstable variable star R Monocerotis, seen at the bottom of this Hubble Space Telescope image. R Mon is a T Tauri star – an unstable stellar newborn that is stll expelling excess material and fluctuating in brightness. Since the nebula is illuminated by reflection of light from R Mon, it varies along with the star, albeit with some delay due to the time light takes to travel along its length.

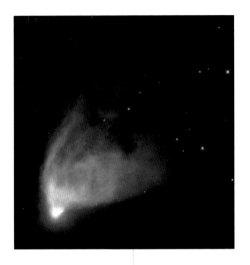

R.A. 06h 39m, DEC. +08º44'
MAGNITUDE 10.0–12.0 (var)
DISTANCE 2,500 light years

Rosette Nebula NGC 2244

This beautiful nebula, to the east of Epsilon Monocerotis, looks like nothing so much as a flower blossoming in space. Officially, NGC 2244 is the star cluster that lies in the middle of all the nebulosity. Packed with heavyweight blue stars, its radiation excites and illuminates the surrounding gas while stellar winds 'hollow out' the central region. NGC 2237, 2238, 2239 and 2246 are the designations for various parts of the Rosette itself. It lies around 5,200 light years from Earth and has a diameter of around 130 light years. While the nebula itself is best spotted using a telescope with low magnification, the star cluster is visible to the naked eye and a good subject for binocular exploration.

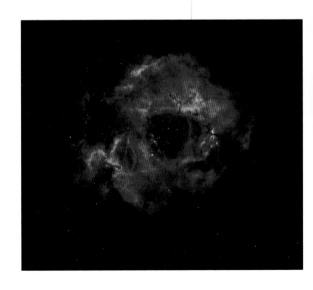

R.A. 06h 34m, DEC. +05º00'
MAGNITUDE 9.0
DISTANCE 5,200 light years

Cone Nebula
NGC 2264

Two related objects share the designation of NGC 2264 – the bright Chrismas Tree star cluster, and a nearby pillar of opaque dust, the Cone Nebula. A nearby emission nebula provides an attractive background to both. The star cluster is easily found with the naked eye or binoculars, and forms a roughly conical shape of its own, which meets the dark cone of gas and dust tip-to-tip. The cone is in fact a 'pillar of creation' – part of a star-forming region that once gave rise to the Christmas Tree Cluster itself, and has now been beaten back to a shadow of its former self by radiation from its offspring.

R.A. o6h 41m, **DEC.** +09°53'
MAGNITUDE 3.9
DISTANCE 2,700 **light years**

Canis Major

Lying to the southeast of Orion, this constellation is unmistakable thanks to the presence of Sirius, the brightest star in the entire sky, which is also known as the Dog Star. However, Canis Major also encompasses other bright stars and deep-sky objects as it is crossed by a rich swathe of the Milky Way.

AT A GLANCE

NAME Canis Major
MEANING The great dog
ABBREVIATION CMa
GENITIVE Canis Majoris
R.A. 06h 50m
DEC. −22° 08'
AREA 380 (43)
BRIGHTEST STAR Sirius (α)

Canis Major was probably first associated with a dog because of the way its stars faithfully pursue the great hunter Orion across the sky. In particular, although Sirius's name derives from the ancient Greek for 'scorching', it has long been popularly known as the 'Dog Star'. This hot white star, with roughly twice the mass of the Sun, is 25 times as luminous, but owes its superlative brightness in our skies largely to its proximity – at a distance of 8.6 light years, it is the fifth-closest star system to Earth.

In contrast, some of the constellation's other bright stars are far more distant, and truly brilliant in their own right – Delta Canis Majoris or Wezen, for example, shines at magnitude 1.8 over a distance of 1,800 light years thanks to an intrinsic luminosity equivalent to around 47,000 Suns.

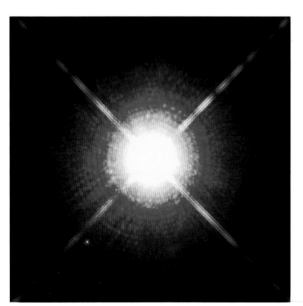

The Sirius System
Despite the Dog Star's proximity to Earth, its binary companion, Sirius B, is tiny and hard to spot. At magnitude 8.5, it is not terribly faint, but its light is simply drowned out by the brilliance of its neighbour. For all its tiny size and faintness, Sirius B has a mass equal to that of the Sun – it is a white dwarf, the burnt-out core of a star that once outshone Sirius A.

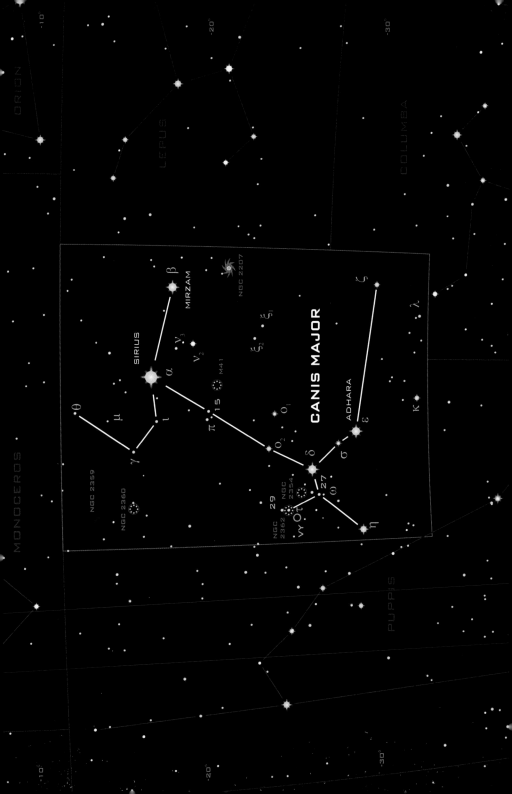

Canis Major Inside View

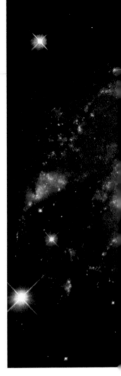

Tau (τ) Canis Majoris Cluster, NGC 2362

This young and heavyweight open cluster surrounds the magnitude-4.4 blue supergiant Tau Canis Majoris. The cluster contains the mass of more than 500 Suns and has a combined apparent magnitude of 4.1, making it an easy naked-eye object and a rewarding sight through binoculars or a small telescope. The Spitzer Space Telescope infrared image (left) was used to search for evidence of protoplanetary discs around these young stars, and revealed that they only persist around the smallest stars in the cluster.

R.A. 07h 19m, DEC. -24°57'
MAGNITUDE 4.1
DISTANCE 5,000 light years

Thor's Helmet
NGC 2359

This beautiful bubble in space is an emission nebula, created by high-energy stellar winds from the blue supergiant star at the bubble's centre. As these particles plough into surrounding material in interstellar space, they create a glowing, roughly spherical shock wave around 30 light years across. As the star moves through space, it carries the bubble with it, and movement through the surrounding nebulosity creates a curving 'bow shock' that sweeps material back around the sides of the nebula and gives it a strong resemblance to a winged helmet.

R.A. 07h 19m, DEC. -13°13'
MAGNITUDE 11.5
DISTANCE 15,000 light years

Supergiant Star
VY Canis Majoris

This red supergiant star just to the south of the Tau Canis Majoris cluster is among the largest stars known. Dropped into the middle of our Solar System, its outer layers would stretch beyond the orbit of Saturn. With 40 times the mass of the Sun, VY is hurtling towards its eventual doom in a supernova explosion. During these final stages of its life it has become an unstable variable, flinging away vast amounts of dust and gas from its outer layers in unpredictable outbursts. This colourful view from the Hubble Space Telescope maps the distribution of dust around the star using polarized light.

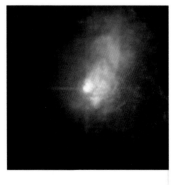

R.A. 07h 22m, DEC. -25°46'
MAGNITUDE 6.5–9.6 (var)
DISTANCE 3,800 light years

Dancing Spirals
NGC 2207/IC 2163

A little way to the south of Beta Canis Majoris lies this attractive pair of galaxies, discovered by English astronomer John Herschel in 1835. The two spirals have been engaged in a slow process of collision for some 40 million years, flinging long streamers of stars out into intergalactic space and triggering bursts of new star formation as their tidal forces tug at each other. Eventually they are doomed to merge, probably forming an elliptical galaxy in a process that will take another 100 million years or more to complete.

R.A. 06h 16m, DEC. -21°22'
MAGNITUDE 12.2, 11.6
DISTANCE 114 million light years

Hydra

AT A GLANCE

NAME Hydra
MEANING The water snake
ABBREVIATION Hya
GENITIVE Hydrae
R.A. 11h 37m
DEC. −14° 32'
AREA 1,303 (1)
BRIGHTEST STAR Alphard (α)

The largest constellation in the sky, Hydra's faint stars wind their way around a huge swathe of the celestial equator to the south of more prominent constellations, such as Leo and Virgo. It was among the 48 constellations listed by the second century AD Graeco-Egyptian astronomer Ptolemy.

This huge and ancient constellation was originally depicted as a serpent by astronomers in ancient Babylon. Although its name recalls the many-headed Lernaean Hydra fought by Hercules, it is more often seen as one player in a story linking the nearby constellations of Corvus and Crater (*see* page 138). Hydra's extensive size and faintness make it hard to identify – only its head and neck, to the southwest of Leo, are reasonably prominent.

The constellation's brightest star, Alphard, is a red giant some 180 light years from Earth. It has a name that means 'solitary one' and, at magnitude 2.0, is indeed the brightest object in a relatively barren region of the sky. Deep-sky objects include the open cluster Messier 48, just below the serpent's head (a good target for binoculars or a small telescope), and the spectacular Southern Pinwheel Galaxy Messier 83 in the constellation's tail.

Galactic Silhouette
NGC 3314, near Hydra's southern boundary, is a rare and beautiful cosmic coincidence – a pair of perfectly aligned galaxies 117 million and 140 million light years from Earth. Light from the more distant galaxy turns the foreground spiral largely transparent, giving us a unique look at its dusty skeleton.

Hydra Inside View

Buckled Spiral
ESO 510-G13

This strange galaxy to the west of Pi Hydrae may only be visible through the largest amateur telescopes, but is fascinating nevertheless. It is clearly a spiral galaxy lying more or less edge-on to Earth, with a dark outer dust lane silhouetted against the background glow of its starry disc and hub, but for some reason the disc has become warped or bent out of shape. Astronomers believe the distortion is caused by gravitational forces at work as ESO 510-G13 swallows up and cannibalizes a smaller galaxy. Intriguingly, there are recent suggestions that our own Milky Way galaxy might have a similar warp on its outer edge.

R.A. 13h 55m, DEC. -26º46'
MAGNITUDE 13.4
DISTANCE 150 million light years

Southern Pinwheel
Messier 83

This beautiful barred spiral is undoubtedly Hydra's most alluring deep-sky object. Roughly one-third the size of the Full Moon but quite bright at magnitude 7.6, it can be located with binoculars on the constellation's southern edge, to the south of R Hydrae in the serpent's tail, and is a rewarding sight through larger instruments. M83 is a barred spiral about half the size of the Milky Way, with prominent dust lanes that can be traced all the way to the core. It was discovered in 1752 by Nicolas Louis de Lacaille, the French astronomer responsible for naming many of the southern constellations.

R.A. 13h 37m, DEC. -29º52'
MAGNITUDE 7.6
DISTANCE 15 million light years

Messier 68

Hydra's only globular cluster lies midway along the line between Beta and Gamma Hydrae. Lying on the opposite side of the sky from the crowded galactic centre, it is among the more isolated globulars in orbit around our galaxy, with an estimated distance of 33,000 light years from Earth.

R.A. 12h 39m, DEC. -28º45'
MAGNITUDE 9.7
DISTANCE 33,000 light years

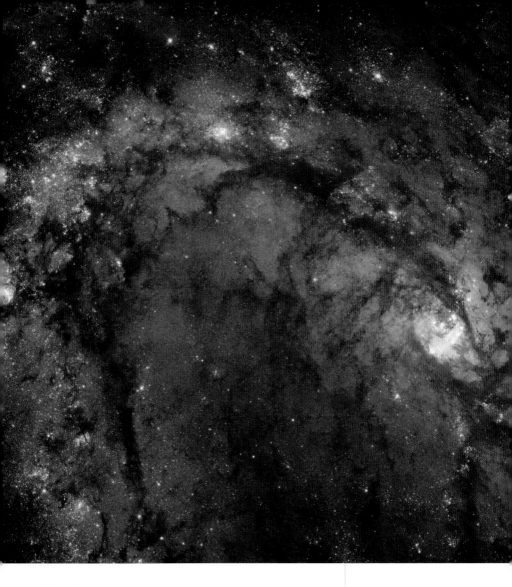

Arc of Starbirth

This stunning image from the Hubble Space Telescope highlights the glowing fires of star formation lighting up one of Messier 83's spiral arms. The galaxy is forming stars at a much faster rate than our own Milky Way, perhaps due to the gravitational influence of nearby galactic neighbours. The image captures key moments in the life cycle of stars, ranging from their birth in dark clouds and bright starbirth nebulae, through their emergence as brilliant open clusters, to their eventual deaths, in the form of some 60 supernova remnants that NASA scientists have counted within the picture.

Corvus, Crater and Sextans

AT A GLANCE

NAME Corvus/Crater/Sextans
MEANING The crow/The cup/The sextant
ABBREVIATION Crv/Crt/Sex
GENITIVE Corvi/Crateris/Sextantis
R.A. 12h 27m/11h 24m/10h 16m
DEC. −18° 26'/−15° 56'/−02° 37'
AREA 184 (70)/282 (53)/314 (47)
BRIGHTEST STAR Gienah (γ)/Delta (δ)/ Alpha (α)

Three small constellations, two of which are ancient and relatively easy to spot, run along the back of the great water snake Hydra, separating it from the zodiac constellations of Virgo and Leo. In mythology, Corvus and Crater are associated with Hydra in a curious story of the god Apollo.

According to this tale, the great god sent his servant the crow to fill his cup with water from a well, but the bird was distracted by a nearby fig tree, and it neglected its task while it waited for the figs to ripen. On the crow's return, it clutched a water snake that it claimed had been blocking the well, but Apollo saw through the ruse, and threw the cup, crow and serpent into the sky in anger. Sextans, by contrast, is a relatively modern and obscure constellation invented by the Polish astronomer Johannes Hevelius in the late 17th century.

Most of the stars in these three constellations are unremarkable, but Delta Corvi (also known as Algorab) is worthy of attention – a small telescope reveals that the magnitude-3.0 primary star has a companion of magnitude 8.5, which often displays an unusual purple colour.

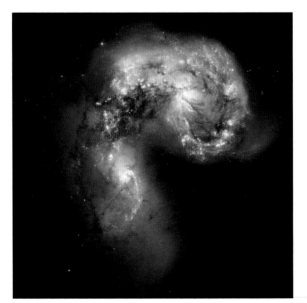

Antennae Galaxies

This strange pair of galaxies lies close to the border of Corvus with Crater, an estimated 45 million light years from Earth. Wide-angle views show long trails of stars resembling an insect's antennae, but detailed images reveal that the Antennae are two spiral galaxies (NGC 4038 and 4039) in the process of collision.

Centaurus

AT A GLANCE

NAME Centaurus
MEANING The centaur
ABBREVIATION Cen
GENITIVE Centauri
R.A. 13h 04m
DEC. −47° 21'
AREA 1,060 (9)
BRIGHTEST STAR Rigil Kentaurus (α)

This sprawling constellation is one of two in the sky that represent mythical centaurs. In Greek mythology centaurs were a race of creatures that were half-human and half-horse. Covering a large area of sky between Hydra and the Southern Cross, it contains several of the highlights of southern skies.

Although it was seen as a bull by ancient Babylonian astronomers, by classical times this group of stars was firmly established as a centaur, and usually associated specifically with Chiron, wise tutor of the warrior Achilles. The centaur's body is formed by a rough pentagon of stars, with the brightest stars of all lying to the southeast.

Alpha Centauri, or Rigil Kentaurus, is the third-brightest star in the sky at magnitude -0.1. Small telescopes reveal that it is a binary consisting of two stars, each very similar to the Sun. An outlying third member of the system, the magnitude-11.0 red dwarf Proxima Centauri, is the closest star to the Sun at a distance of 4.26 light years. Beta Centauri (Hadar) is also a triple star system, but in this case one that consists of heavyweight and highly luminous blue stars some 525 light years from Earth (only two of which can be visually separated).

Stars of the Centaur

This stunning view of the southern Milky Way shows the extensive constellation of Centaurus. Alpha and Beta Centauri form a spectacular pair at lower left, alongside the Southern Cross. In 2012, astronomers announced the discovery of an Earth-sized planet orbiting the fainter of Alpha Centauri's near-twin stars.

Centaurus Centaurus A/NGC 5128

Radio Jets

The bright galaxy NGC 5128, discovered in 1826 by James Dunlop, is easily spotted with binoculars and appears to be an elliptical ball of stars through small telescopes. Slightly larger instruments or long-exposure photographs will reveal that this bright disc is bisected by a dark stream of dust. NGC 5128 coincides with a strong radio source known as Centaurus A, and as a result the galaxy itself is informally known by this name as well. Centaurus A is one of the closest and brightest active galaxies to Earth – multiwavelength composite images such as the one reproduced right reveal twin jets of material emerging in opposite directions from the bright nucleus, emitting both radio waves (shown in orange) and X-rays (in blue).

R.A. 13h 26m, DEC. -43°01'
MAGNITUDE 7.0
DISTANCE 15 million light years

Cannibal Galaxy

This unusual image combines a near-infrared view of Centaurus A (background) with a map derived from radio observations of carbon monoxide molecules within the galaxy. It reveals an almost perfect trapezium of gas-rich material embedded within the galaxy and coinciding with the visible dust lane. Different colours on opposite sides of the nucleus show regions that are moving towards Earth (green) or away (red) – evidence that the entire structure is rotating. This strange feature is the surviving skeleton of a gas- and dust-rich galaxy (probably a spiral similar to our own Milky Way) that has been engulfed by a largely gasless elliptical. The blaze of light from the centre of the infrared image indicates the precise position of the galaxy's active nucleus, a supermassive black hole that has been stirred into life by these tumultuous events.

Starburst Region

Elliptical galaxies normally lack the raw material to fuel ongoing star formation, and so tend to be dominated by long-lived, lower-mass stars. The injection of new material into Centaurus A, however, has triggered an enormous new wave of starbirth around and within the dark band. This Hubble Space Telescope image reveals the telltale pinkish glow of huge starbirth nebulae, and the brilliant blue of young star clusters emerging from their cocoons. The multiwavelength view combines visible light with near-infrared and ultraviolet images to render much of the material in the galaxy's dust lane transparent.

Centaurus Omega Centauri NGC 5139

Omega Centauri

Bright enough to be given its own Greek 'star letter', Omega Centauri is the largest and brightest globular cluster in the heavens. Easily visible to the naked eye as a patch of light about the size of the Full Moon, it shines at an impressive magnitude 3.7. In part, this is due to its proximity – at 16,000 light years it is one of the closer globulars to Earth. But Omega is also a genuinely impressive object in its own right, an enormous stellar conglomerate of several million stars crowded into a region just 170 light years across. Any optical instrument will begin to show individual stars in the cluster's outer edges, but only the largest instruments can resolve the blizzard of stars near its centre.

R.A. 13h 27m, DEC. -47º29'
MAGNITUDE 3.7
DISTANCE 16,000 light years

A Sky Full of Stars

This Hubble Space Telescope view peers into the crowded heart of Omega Centauri, capturing tens of thousands of stars in a single image. Stars within globular clusters are extremely old – they formed in a much earlier phase of cosmic evolution and are only so long-lived because they lack metal 'pollutants' – heavier elements that accelerate the rate at which more recently formed stars burn their fuel. Evidence suggests Omega's stars formed in several stages over some 2 billion years, concluding 10 billion years ago. In contrast, most globulars seem to have formed in a single burst. This lends weight to a theory that Omega is not a normal globular at all, but is actually the dense core of a dwarf galaxy cannibalized by the Milky Way.

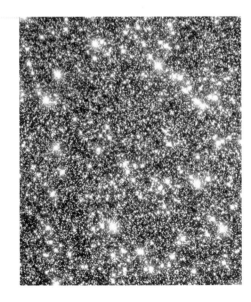

Stellar Variety

This detailed view from Hubble's Wide Field Camera 3 shows a colourful array of stars in the heart of Omega Centauri. Yellow stars, representing the vast majority of the cluster's population, are still in the midst of their prolonged, sedate middle age. Red stars are giants that have swollen and brightened as they near the end of their lives, while faint blue-white dots are white dwarfs – the burnt-out cores of stars that have long since died. The bright blue stars are known as 'blue stragglers'. In theory, there is no place for short-lived heavyweight stars in such a cluster – astronomers think they probably formed relatively recently through the collision and merger of 'normal' stars in the cluster's crowded heart.

Lupus

This bright but somewhat shapeless constellation is best located by looking between Antares in Scorpius to its north, and the bright pair of Alpha and Beta Centauri to the south. Lying across a bright region of the Milky Way, it contains a variety of interesting celestial objects.

AT A GLANCE

NAME Lupus
MEANING The wolf
ABBREVIATION Lup
GENITIVE Lupi
R.A. 15h 13m
DEC. −42° 43'
AREA 334 (46)
BRIGHTEST STAR Alpha (α)

Although this star grouping dates back at least to classical Greece, it was not identified as a wolf until medieval times. Earlier astronomers knew it as Bestia or Fera, both words that simply mean 'wild animal'. The beast was often shown impaled on a spear wielded by neighbouring Centaurus.

The constellation's brightest star, Alpha Lupi or Kakkab, is a blue stellar giant shining at magnitude 2.3 across 550 light years. Mu Lupi, meanwhile, is an attractive multiple star – small telescopes reveal a magnitude-7.2 companion in orbit around a magnitude-4.3 primary, while larger instruments split the primary into twin components of magnitudes 5.1 and 5.2. Numerous star clusters are scattered through the constellation, with NGC 5822 the brightest of the open clusters, and NGC 5986, at magnitude 7.1, the brightest of several globulars.

Retina Nebula
Catalogued as IC 4406, this unusual, apparently rectangular cloud of gas is in fact a planetary nebula – an expanding shell of gas puffed out into space from the outer layers of a dying Sun-like star. The Retina's unique shape is due to the particular angle at which we see it – in reality, the nebula is roughly doughnut-shaped, but we are viewing it from almost exactly edge-on.

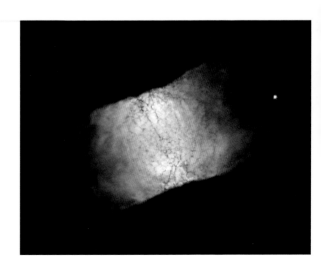

Scorpius

AT A GLANCE

NAME Scorpius
MEANING The scorpion
ABBREVIATION Sco
GENITIVE Scorpii
R.A. 16h 53m
DEC. −27° 02'
AREA 497 (33)
BRIGHTEST STAR Antares (α)

This large and bright zodiac constellation, crossing the southern Milky Way between Sagittarius and Libra, is easily identified thanks to the presence of blood-red Antares, the so-called 'rival of Mars'. This is a red supergiant star and the sixteenth brightest star in the nighttime sky.

Scorpius is one of the oldest constellations, and references to it are found in even the earliest astronomical records. While its twisting form does indeed bear some resemblance to the scorpion that gives it its name, its modern pattern is truncated at the head, with the claws forming nearby Libra.

Shining at an average of magnitude 1.0, Antares is subject to slow variations in its light and is one of the most extreme stars known – dropped into the middle of the Solar System, it would engulf all the planets out to Jupiter. Antares has a close companion of magnitude 5.5 visible in medium-sized telecopes, but the constellation has some easier multiple targets. Acrab (Beta Scorpii) reveals blue components of magnitudes 2.6 and 4.9 through small telescopes, while Jabbah (Nu Scorpii), is a 'double double' that reveals two components when viewed with a small telescope, and four with a medium-sized instrument.

Butterfly Cluster
This contrasting star cluster in eastern Scorpius is visible to the naked eye as a concentrated patch of light in the Milky Way, roughly the size of the Full Moon. Catalogued as Messier 6, it lies around 1,600 light years from Earth. Though dominated by hot blue stars, its brightest star is the orange giant BM Scorpii.

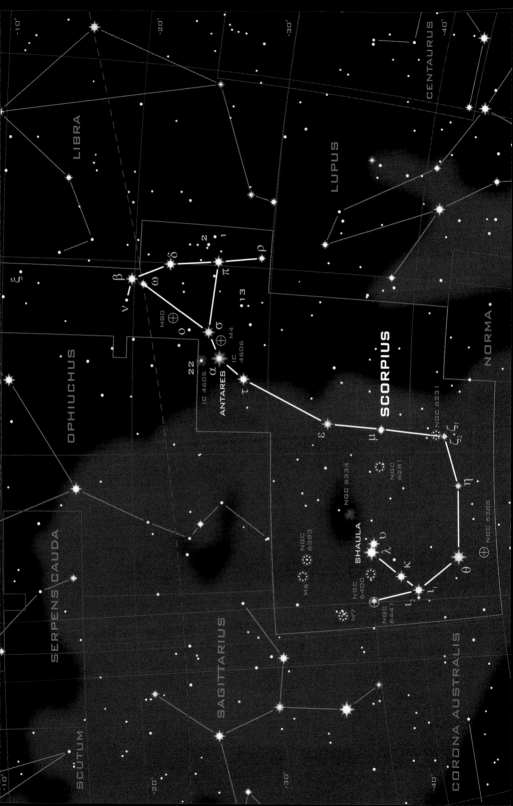

Scorpius Inside View

Messier 80

This globular cluster is among the densest known, containing hundreds of thousands of stars in a ball with a diameter of just 95 light years. It is easily located with binoculars or a small telescope, midway between Antares and Beta Scorpii, but even the largest instruments will be unable to resolve detail in its crowded core. Perhaps unsurprisingly for such a dense cluster, M80 contains a large number of 'blue stragglers', unexpectedly bright and hot stars. These are thought to form through collisions and mergers among the more sedate stars of a cluster's general population.

R.A. 16h 17m,
DEC. -22°59'
MAGNITUDE 7.9
DISTANCE 32,600
light years

Ptolemy Cluster
Messier 7

This knot of bright stars in the Milky Way close to the scorpion's sting is named after the Greek-Egyptian astronomer Ptolemy of Alexandria, who wrote the first description of it around AD 130. Naked-eye stargazers may have difficulty picking it out against the background star clouds of this region, but binoculars or a low-powered telescope will show the cluster in all its glory. Unlike M4 and M80, Messier 7 is an open cluster. Its loose collection of several dozen stars formed around 200 million years ago and is dominated by bright blue stars and a single red giant.

R.A. 17h 54m,
DEC. -34°49'
MAGNITUDE 3.3
DISTANCE 800 light
years

Messier 4

In contrast to M80, the globular cluster Messier 4 shows a relatively loose structure, as displayed in this Hubble Space Telescope view of its central regions. The cluster is on the edge of naked-eye visibility, but is easily spotted with binoculars or a small telescope, lying directly to the west of Antares. Slightly larger instruments will begin to resolve individual stars across the cluster's 75-light-year diameter. At a distance of just 7,200 light years, M4 is one of the closest globular clusters to Earth. Studies of the cluster have identified white dwarfs within it whose lives began around 13 billion years ago – making them the oldest stars associated with the Milky Way.

R.A. 16h 24m, DEC. -26°32'
MAGNITUDE 5.9
DISTANCE 7,200 light years

Antares
Alpha (α) Scorpii

Brilliant red Antares is normally the sixteenth-brightest star in the sky, a highly evolved red supergiant that varies slowly and erratically between magnitudes 0.9 and 1.2. It is the brightest member of the Scorpius–Centaurus OB Association, a nearby group of massive stars, and is also the group's heaviest member. With a mass of 17 Suns, Antares has squandered its supplies of nuclear fuel at a tremendous rate, and is now hurtling towards destruction, despite being a mere 12 million years old. As this stunning image shows, Antares lies within a dense region of nebulosity, impressively revealed in long-exposure photographs.

R.A. 16h 29m,
DEC. -26º26'
MAGNITUDE 0.9–1.2 (var)
DISTANCE 600 light years

Sagittarius

Lying in one of the richest regions of the Milky Way, this constellation is strewn with numerous deep-sky objects ranging from nebulae to star clusters and the very centre of our galaxy itself. The brightest stars of Sagittarius form a distinctive 'teapot' shape in front of dense Milky Way starfields.

AT A GLANCE

NAME Sagittarius
MEANING The archer
ABBREVIATION Sgr
GENITIVE Sagittarii
R.A. 19h 06m
DEC. −28° 29'
AREA 867 (15)
BRIGHTEST STAR Kaus Australis (ε)

Several ancient civilizations saw it as a horse and rider, but the ancient Greeks identified Sagittarius as a centaur (half-man, half-horse) wielding a bow and arrow. The brightest star is Epsilon Sagittarii (Kaus Australis), a 'white giant' star of magnitude 1.8 some 145 light years from Earth. Beta Sagittarii or Arkab, meanwhile, is a line-of-sight double consisting of blue and white stars at magnitudes 4.0 and 4.3, 380 and 140 light years away respectively. A small telescope will show a third component of magnitude 7.1.

Deep-sky highlights within Sagittarius include the Lagoon, Omega and Trifid nebulae (M8, M17 and M20). The heart of the Milky Way, marked by the radio source Sagittarius A*, lies 26,000 light years away close to Gamma Sagittarii, but is hidden behind dense intervening star clouds.

Trifid Nebula
Dark canyons of dust divide this glowing gas cloud into three roughly equal parts and give the Trifid Nebula, Messier 20, its name. It lies around 5,200 light years from Earth and presents a rare and beautiful combination of a pinkish emission nebula, a blue reflection nebula and a star cluster that provides the radiation to illuminate them both, glowing at just below naked-eye visibility.

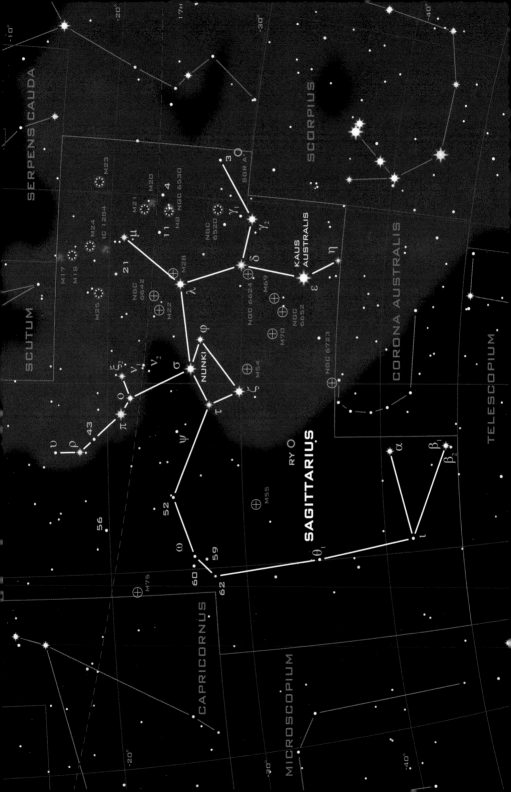

Sagittarius Inside View

Alien Visitor
Messier 54

At first glance this globular cluster seems to be a typical example of its type, but Messier 54 hides an extraordinary secret. Although the cluster has been recognized since Charles Messier first catalogued it in 1778, it was only in 1994 that astronomers analysing the density of star clouds within and beyond

the galactic centre identified the ghostly outline of a small galaxy in the process of colliding with our own. Recognition of this galaxy, the Sagittarius Dwarf Elliptical or SagDEG, led to the realization that M54 is associated with it – this is an extragalactic cluster in the process of being captured by the Milky Way.

R.A. 18h 55m, DEC. -30º29'
MAGNITUDE 8.4
DISTANCE 87,000 light years

Omega Nebula
Messier 17

Also known as the Swan Nebula, Messier 17 is best located with binoculars, sweeping down the length of nearby Scutum to the border with Sagittarius looking for a wedge-shaped patch of nebulosity. Through larger telescopes, the nebula takes on a structure similar to the Greek capital Omega (W). M17 is one of the brightest and most intense star-forming regions in the Milky Way, and would be a far more spectacular object if it was closer to Earth. As it is, it still offers some rewarding views for stargazers.

R.A. 18h 20m, DEC. -16º11'
MAGNITUDE 6.0
DISTANCE c.5,500 light years

Lagoon NebuLa
Messier 8

Standing out from the Milky Way under the darkest skies, this cloud of delicately glowing gas in western Sagittarius is one of only two star-forming nebulae visible to naked-eye observers from the northern hemisphere. Binoculars will show a bright nucleus at the heart of a glowing cloud spread across the width of three Full Moons, while small telescopes will begin to reveal some of its detail including the bisecting stream of dust that gives the nebula its name. A naked-eye star cluster, NGC 6530, overlays the nebula's eastern half and is thought to have emerged in the last 2 million years.

R.A. 18h 04m, DEC. -24º23'
MAGNITUDE 6.0
DISTANCE 5,000 light years

Sagittarius The Galactic Centre

High Contrast

This beautiful long-exposure view of the region around Sagittarius and Scorpius captures the blaze of light from countless stars in the Milky Way, contrasting with the spine of dust than runs along the Sagittarius/ Carina spiral arm. The bright knots of the Lagoon and Trifid nebulae are on the extreme left, while Antares and the Rho Ophiuchi Complex dominate the right-hand side of the image. The small square marks the location of the Sagittarius A* radio source at the centre of the Milky Way, 20,000 light years beyond most structures visible in this picture. Invisible radiations like radio and X-rays can see through intervening material to lift the veil on our galaxy's central regions.

R.A. 17h 46m, DEC. -29⁰00'
MAGNITUDE N/A
DISTANCE 26,000 light years

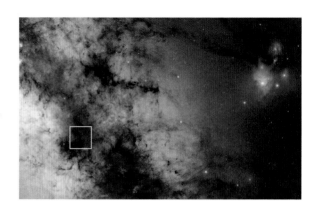

Multiwavelength Milky Way

This colourful image combines views of the Milky Way's central region taken by three of NASA's 'Great Observatories' satellites. Near-infrared light captured by the Hubble Space Telescope is shown in yellow, a deep-infrared view from the Spitzer Space Telescope appears in red, while data from the Chandra X-Ray Observatory is coded blue. Lifting the veil on the heart of our galaxy reveals a landscape of twisted nebulae and dust clouds, shaped by violent stellar winds, supernovae shock waves and powerful gravitational forces. Sagittarius A*, at the heart of our galaxy, is just to the right of centre.

Sagittarius A*

This image from the Chandra X-Ray Observatory is the highest-resolution view of the Milky Way's central region yet obtained. Amid the clouds of reddish lower-energy radiations, lobes of gas emitting powerful X-rays extend on all sides of the Sagittarius A* object. Hubble Space Telescope observations have revealed the presence of clusters of giant stars surrounding Sagittarius A*, orbiting under the influence of an object that contains 4 million solar masses of material in a region of space smaller than the orbit of Uranus. The fact that there is no visible object coinciding with the radio source and the cluster's centre of mass is clinching evidence for a supermassive black hole acting as the gravitational 'anchor' for our entire galaxy.

Capricornus

This curious zodiac constellation to the east of Sagittarius once marked the spot where the Sun reached its annual southerly extreme. Rather faint and shapeless, it has only one star brighter than magnitude 3.0. It is the smallest constellation in the zodiac and the second faintest after Cancer.

Despite its disadvantages, Capricornus is one of the most ancient constellations – Babylonian and Assyrian astronomers referred to it as the 'goat-fish', while other cultures saw it as a more normal goat, ibex or ox. Greek astronomers, drawing on the Mesopotamian form, saw the constellation as a depiction of the goat-headed god Pan, transforming into a fish to escape from the monstrous Typhon.

The brightest star in this region of the sky is Delta Capricorni. This eclipsing binary system normally shines at magnitude 2.9, but briefly drops by 0.2 magnitudes once every 24.5 hours as its tightly orbiting components pass in front of each other. Alpha Capricorni (Algiedi), meanwhile, is a naked-eye double with two stars of magnitudes 3.6 and 4.2 at wildly different distances (108 and 635 light years). A small telescope will reveal that each of these stars has a fainter binary companion.

Captured Cluster
Globular cluster Messier 30 is the deep-sky highlight of Capricornus – a dense cloud of stars just over 29,000 light years from Earth, easily detectable through binoculars at magnitude 7.7. Telescopes will reveal that it has undergone a process known as 'core collapse', in which large numbers of stars fall into the centre of the cluster.

Piscis Austrinus and Microscopium

AT A GLANCE

NAME Piscis Austrinus/Microscopium
MEANING The southern fish/
  The microscope
ABBREVIATION PsA/Mic
GENITIVE Pisces Austrini/Microscopii
R.A. 22h 17m/20h 58m
DEC. −30° 39'/−36° 16'
AREA 245 (60)/210 (66)
BRIGHTEST STAR Fomalhaut (α)/Gamma (γ)

The southern fish is among the most southerly of the ancient Greek constellations, and is easily identifiable thanks to the presence of the brilliant star Fomalhaut. The more recently invented Microscopium is far more obscure.

In mythological terms, the southern fish was sometimes depicted as parent to the more famous fish in Pisces, drinking from the waters that poured from the cup of Aquarius. Microscopium, in contrast, is one of several more modern constellations invented by French astronomer Nicolas Louis de Lacaille in the 1750s.

Fomalhaut, the brightest star in this region of the sky at magnitude 1.2, is relatively close to Earth at a distance of 25 light years. A relatively young star at around 300 million years old, it is today celebrated for the disc of protoplanetary material (and possible planets) that orbit around it. Beta and Gamma Piscis Austrini, meanwhile, are double stars of middling brightness, which require a small or medium-sized telescope respectively to reveal their fainter companions.

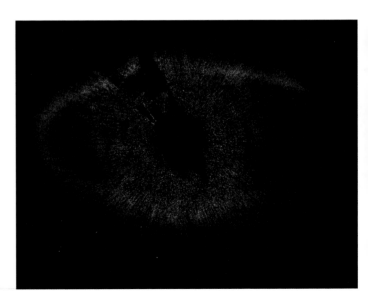

Eye of God
This extraordinary image from the Hubble Space Telescope reveals fine structure in the ring of material orbiting Fomalhaut (the star itself lies in the blanked-out area at the centre of the picture). The shape of this 'debris disc', and in particular the sharp inner edge of the ring, suggests the presence of an unseen planet with at least the mass of Neptune.

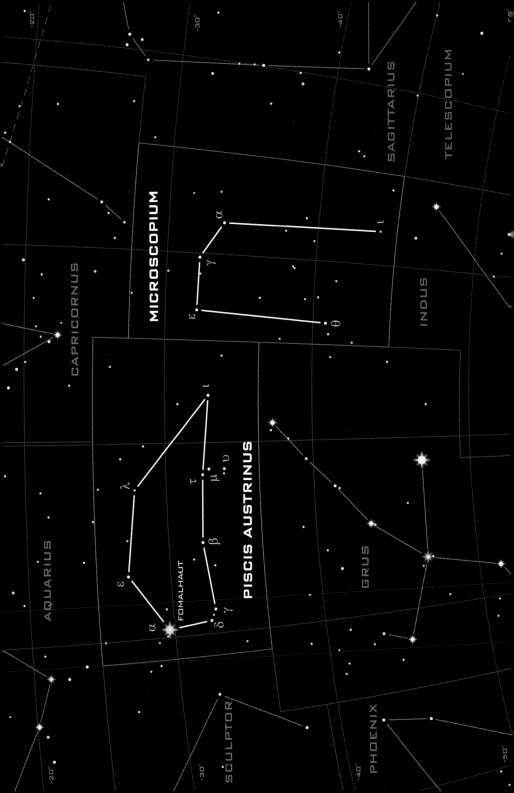

Sculptor and Fornax

These two obscure constellations lie to the south of Cetus, tucked into a bend of the celestial river Eridanus. Sculptor is imagined as representing a sculptor's workshop while Fornax is a type of furnace. Both are faint and obscure, but by coincidence each is home to a significant cluster of galaxies.

AT A GLANCE

NAME Sculptor/Fornax
MEANING The sculptor/The furnace
ABBREVIATION Scl/For
GENITIVE Sculptoris/Fornacis
R.A. 00h 26m/02h 48m
DEC. −32° 05'/−31° 38'
AREA 475 (36)/398 (41)
BRIGHTEST STAR Alpha (α)/Alpha (α)

Fornax and Sculptor are both reasonably typical of the constellations added to southern skies by the French astronomer Nicolas Louis de Lacaille in his 1763 atlas of southern stars. Lacaille was an assiduous observer, cataloguing the positions of some 10,000 stars over four years working at the Cape of Good Hope from 1750, but his constellations are mostly faint and difficult to identify.

Sculptor lies in the direction of our galaxy's south pole, and offers clear views of intergalactic space. Several nearby galaxies – the closest galaxy group to our own – lie in this direction, while Fornax is host to the far richer and more distant Fornax Cluster. Alpha Fornacis is a binary some 42 light years from Earth, consisting of yellow stars of magnitudes 4.0 and 6.5 that can be split with binoculars or a small telescope. Alpha Sculptoris is a blue giant of magnitude 4.3, 780 light years away.

Fornax Cluster

Fornax is home to a rich cluster of galaxies – the second most impressive in our cosmic neighbourhood after the Virgo Cluster (see page 92). The Fornax Cluster is centred about 65 million light years from Earth, and is probably linked with another cluster in Eridanus, 20 million light years further away. The core of the cluster occupies a region around 2 degrees across, concentrated around the giant galaxies NGC 1316 and NGC 1365, and is a good target for small telescopes.

Sculptor and Fornax Inside View

Silver Coin Galaxy
NGC 253

Midway between Alpha Sculptoris and Beta Ceti lies one of the brightest galaxies in southern hemisphere skies – a nearby spiral that we view from almost edge-on. NGC 253 is the dominant member of the Sculptor Group, the nearest galaxy group beyond our own Local Group. The galaxy's narrow profile concentrates its light and makes it easy to spot with binoculars, while larger telescopes will reveal a mottled ellipse with hints of spiral structure. NGC 253's spiral arms are unusually bright in comparison to its core, indicating that it is currently undergoing a major burst of star formation.

R.A. oh 48m,
DEC. -25°17'
MAGNITUDE 7.1
DISTANCE 10 million
light years

Ringed Galaxy
NGC 1097

The bright central regions of this barred spiral in Fornax are visible through a small telescope to the northwest of Beta Fornacis, but larger telescopes are needed to reveal its complete structure. NGC 1097 is currently undergoing a wave of star formation that lights up its spiral arms. The activity may be connected to an ongoing close encounter with the small elliptical galaxy NGC 1097a (at upper left in this picture). One of the galaxy's most unusual features is a perfect ring of star formation surrounding the core, and this is almost certainly linked to the current galactic interaction as well.

R.A. o2h 46m, **DEC.** -30°17'
MAGNITUDE 10.2
DISTANCE 45 million light years

Borderline Case
NGC 55

This peculiar galaxy, displaying characteristics somewhere between barred spiral and irregular galaxies, lies on the southern border of Sculptor, a little to the northwest of Alpha Phoenicis. Although it lies close to the other galaxies of the Sculptor Group, and has traditionally been seen as a member of that group, recent studies suggest it may be an outlying member of our own Local Group, or even an independent galaxy with no gravitational allegiance to either group. Its distortion is almost certainly due to the proximity of the small spiral galaxy NGC 300.

R.A. oh 15m,
DEC. -39°11'
MAGNITUDE 8.8
DISTANCE 7 million
light years

Hubble
Ultra-Deep Field

Inspired by the success of the original 'Hubble Deep Field' observations (*see* page 34), astronomers turned their attention in 2003 to an even more ambitious project. With a combined exposure time of a million seconds, the Ultra-Deep Field (HUDF) combines visible-light observations with near-infrared data from Hubble's NICMOS camera, revealing objects so distant (and therefore receding rapidly due to the expansion of the Universe) that their light has been Doppler-shifted into the infrared. The result is a view that looks back through time to a mere 400 million years after the Big Bang.

R.A. 03h 33m, DEC. -27º47'
MAGNITUDE < 29.0
DISTANCE Up to 13.3 billion light years

Eridanus

AT A GLANCE

NAME Eridanus
MEANING The river
ABBREVIATION Eri
GENITIVE Eridani
R.A. 03h 18m
DEC. −28° 45'
AREA 1,138 (6)
BRIGHTEST STAR Achernar (α)

The twisting constellation of the celestial river winds its way back and forth across the sky, trending southwards from its origin by Orion's foot, to its end point at brilliant Achernar (also known as Alpha Eridani). This star spins at a speed of more than 250 km (155 miles) per second.

Greek astronomers associated Eridanus with a variety of rivers both real and mythological. However, the ancient constellation ended at the star now designated Theta Eridani. The river was only extended to the south after Renaissance European explorers reported the presence of the bright star that lay out of sight to contemporary observers from Mediterranean latitudes.

Achernar, whose name derives from the Arabic for 'river's end', lies some 143 light years from Earth and is a hot blue-white star shining at magnitude 0.5. It is also one of the fastest-rotating stars in the sky, spinning so rapidly that its equator bulges by 50 per cent compared to its poles. Other notable stars include the Sun-like Epsilon Eridani, and the double star Omicron 2 Eridani. A small telescope will reveal that the magnitude-4.4 primary has a companion of magnitude 9.5 – the most easily seen white dwarf in the sky.

Epsilon Eridani

Officially the third-closest naked-eye star system to the Sun, Epsilon Eridani shines at magnitude 3.7 and lies 10.5 light years away. It has the mass of 0.8 Suns and is relatively young at around a billion years old. An excess of infrared radiation was the first hint that Epsilon was surrounded by a disc of potentially planet-forming dust. In 2000, the presence of a giant planet in a seven-year orbit around the star was confirmed, and the system is now known to contain two rocky asteroid belts and possibly at least one more planet.

Lepus

Directly to the south of Orion lies a bow-tie-shaped group of fainter stars representing a hare. Four stars of this constellation form a quadrilateral and are known as 'the Throne of Jawza'. Although it is lacking in impressive deep-sky objects, this constellation is home to some intriguing stars.

Lepus is normally associated with a hare being chased by Orion's dogs, and is usually shown crouching unnoticed at the hunter's feet as Orion himself turns to face the charging bull Taurus. Some Arab astronomers, however, saw it as a pair of camels drinking from the sweet waters of the celestial river Eridanus.

Alpha Leporum, or Arneb, is a rare white supergiant of magnitude 2.6, around 1,300 light years from Earth and 13,000 times more luminous than the Sun. Gamma Leporis, meanwhile, is an attractive double star with yellow and orange components of magnitudes 3.6 and 6.2, divisible through binoculars or a small telescope. South of the main star pattern, meanwhile, lies Messier 79, a globular cluster of magnitude 8.6 some 41,000 light years from Earth and best seen with a small telescope.

Hind's Crimson Star

Close to the western boundary of Lepus, binocular observers can find R Leporis, one of the reddest stars in the sky. Also known as Hind's Crimson Star, it lies some 1,500 light years from Earth and is a long-period variable similar to Mira in Cetus (see page 118). It varies slowly in size and brightness, usually between magnitudes 7.3 and 9.8, with a period of around 420 days. However, on occasion it can become visible to the naked eye, reaching magnitude 5.5.

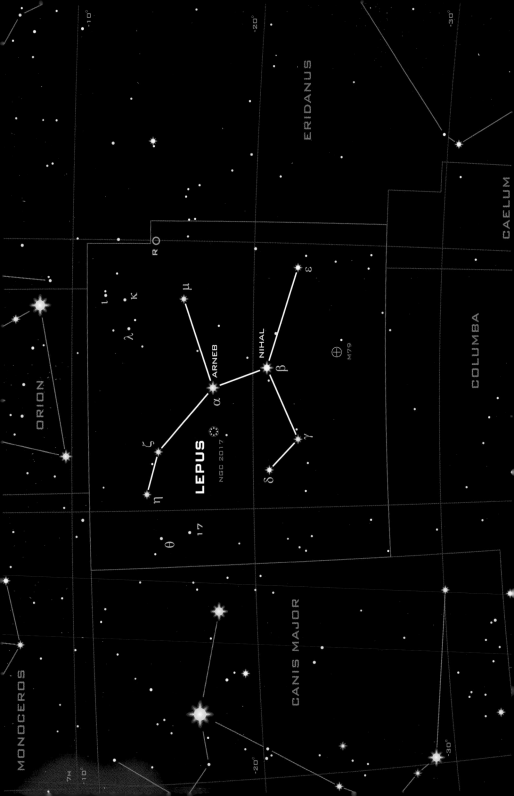

Caelum and Columba

Two small constellations lie to the south of Lepus, between the equatorial stars of Orion and the equally prominent southern stars of Vela and Carina. These form the relatively bright constellation of the dove (Columba) and the more obscure chisel (Caelum) which is the eighth-smallest constellation.

AT A GLANCE

NAME Caelum/Columba
MEANING The chisel/The dove
ABBREVIATION Cae/Col
GENITIVE Caeli/Columbae
R.A. 04h 42m/05h 52m
DEC. −37° 53'/−35° 06'
AREA 125 (81)/270 (54)
BRIGHTEST STAR Alpha (α)/Phact (α)

Columba owes its origins to Dutch astronomer and theologian Petrus Plancius, who inserted several biblically inspired constellations into a celestial chart published in 1592 – he intended it to represent the dove sent out by Noah from the Ark. Caelum, meanwhile, is an invention of French astronomer Nicolas Louis de Lacaille from the 1750s.

Alpha Columbae, or Phact, is a fast-spinning blue-white star of average magnitude 2.6 around 530 light years from Earth. As it spins, it flings out material from around its equator, creating a series of rings that cause its brightness to fluctuate slightly. Another interesting star is Mu Columbae, a hot blue stellar runaway of magnitude 5.1, around 1,300 light years from Earth. Its fast-moving track through space, like that of AE Aurigae (*see* page 28), can be traced back to the Orion Nebula.

Mixed Globular
NGC 1851, in the southwestern corner of Columba, is one of the sky's more distant globular clusters, at 39,500 light years from Earth. Despite this, it is relatively bright at magnitude 7.3 and can usually be spotted through good binoculars. NGC 1851 displays several unusual features that suggest it may have formed from the merger of two separate globular clusters at some time in its history.

Puppis

AT A GLANCE

NAME Puppis
MEANING The stern of Argo
ABBREVIATION Pup
GENITIVE Puppis
R.A. 07h 15m
DEC. −31° 11'
AREA 673 (20)
BRIGHTEST STAR Naos (ξ)

The northernmost part of Argo, the great celestial ship, rises into equatorial skies to the south and east of Canis Major. Crossed by a broad swathe of the Milky Way, Puppis is rich in dense star clouds and clusters. Several extrasolar planetary systems have also been found around stars in this constellation.

Puppis marks the stern of the mighty ship Argo. Ancient Greek stargazers recognized this huge constellation sailing across their southern horizon during the northern hemisphere spring, but later astronomers split it into three separate parts, the stern (Puppis), keel (Carina) and sail (Vela).

One peculiar consequence of Argo's dissolution is that the three modern constellations share one set of Greek 'Bayer letters' between them – hence Naos, this constellation's brightest star, is designated Zeta Puppis. This extremely hot blue star of magnitude 2.2 is just 4 million years old and lies roughly 1,100 light years away. Another highlight is L Puppis, a line-of-sight double star easily separated with binoculars. The northern component, blue-white L1 Puppis, shines at a steady magnitude 4.9, but its southern companion, the reddish L2, varies between magnitude 2.6 and 6.2 in a 141-day cycle.

Cauldron of Starbirth

The seething and bubbling gases revealed within nebula NGC 2467 by this Hubble Space Telescope image have been compared to a witch's cauldron. The nebula lies close to Omicron Puppis and forms the background to a star cluster of magnitude 7.1 that can be located with binoculars. Despite appearances, the cluster's stars are thought to lie some way in front of the nebula.

Antlia and Pyxis

These two obscure constellations of mid-southern skies are best found by looking in the empty region west of Centaurus and east of Puppis, northwards of the bright stars of Vela. Pyxis and Antlia are both 18th-century inventions of the French astronomer Nicolas de Lacaille.

AT A GLANCE

NAME Antlia/Pyxis
MEANING The air pump/The ship's compass
ABBREVIATION Ant/Pyx
GENITIVE Antliae/Pyxidis
R.A. 10h 16m/08h 57m
DEC. −32° 29'/−27° 21'
AREA 239 (62)/221 (65)
BRIGHTEST STAR Alpha (α)/Alpha (α)

Typically neither of them are particularly convincing as star patterns. Like most of Lacaille's constellations, they represent scientific instruments of the time – in this case a magnetic compass (Pyxis) and a pneumatic air pump (Antlia) that was invented by the French physicist Denis Papin in the late 17th century.

Zeta Antliae is a deceptive double – binoculars reveal a pair of white stars of magnitudes 5.8 and 5.9, but even though they are at roughly the same distance (around 370 light years), they are not bound in orbit around each other, and so are technically a line-of-sight double. Small telescopes show that the western component (Zeta 1) is a true binary of magnitudes 6.2 and 7.0. T Pyxidis, meanwhile, is a recurrent nova – a normally obscure system of magnitude 13.8, some 6,000 light years from Earth, that flares into binocular or even naked-eye visibility every two or three decades (this happened most recently in 1996).

Tilted Spiral
NGC 2997 is a bright spiral galaxy to the southwest of Theta Antliae, presented at an angle of roughly 45 degrees. It lies at a distance of around 40 million light years from Earth and forms the gravitational anchor for its own small galaxy group. With a diameter roughly one-third that of the Full Moon and a magnitude of 10.1, it is a good challenge for locating with smaller telescopes, though much larger instruments are needed to bring out its detail.

Vela

AT A GLANCE

NAME Vela
MEANING The sails of Argo
ABBREVIATION Vel
GENITIVE Velorum
R.A. 09h 35m
DEC. −47° 10'
AREA 500 (32)
BRIGHTEST STAR Regor (γ)

The irregular octagonal shape of this constellation is not the most obvious of sky patterns, but it is identifiable thanks to its brightness and location between Centaurus and Crux to the east, and Carina to the west. The largest known emission nebula, the Gum Nebula, is found here and extending into Puppis.

Vela marks the sail of the ship Argo, an ungainly constellation that has since been broken into three segments. In mythology, Argo was built for Jason by the great shipwright Argus, and placed under the protection of Hera, the Queen of the Gods.

The constellation's brightest star, Gamma Velorum, or Regor, is a complex multiple star of magnitude 1.8. Binoculars reveal one companion star of magnitude 4.3, while small telescopes will show two more of magnitudes 8.5 and 9.4. Studies of the spectrum of Gamma 1's light reveal that it is also a binary star with components 10 and 30 times the mass of the Sun. Delta Velorum is another multiple, its most obvious components a white star of magnitude 2.0, and a yellow one of magnitude 5.1. Another attractive object is IC 2391, a naked-eye open cluster centred on the star Omicron Velorum.

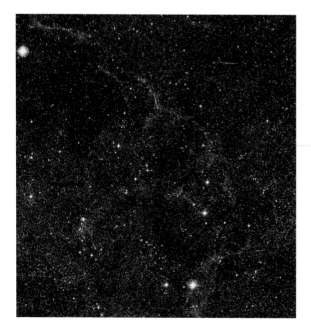

Vela Supernova Remnant

Northeast of Gamma Velorum, a network of faintly glowing gaseous tendrils marks the outline of the Vela Supernova Remnant (SNR) – an expanding shockwave of hot gas generated by a stellar explosion around 11,000 years ago. For visual observers, the Vela SNR is frustratingly faint, and best seen through long-exposure photographs.

Vela Inside View

HH 47

Herbig Haro Objects are twin-lobed nebulae found on either side of young stars and often connected to them by narrow jets. They are formed when a rapidly spinning newborn star flings off excess material

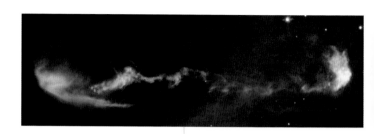

along its axis of rotation. As these high-speed jets stream out into space, they may encounter interstellar gas and dust, creating a glowing shock wave. Herbig Haro 47 (HH 47) is perhaps the best known and most studied example of these intriguing objects, with jets that display numerous complex knots throughout their length. The Hubble Space Telescope image above shows the shock front on one side of the nebula.

R.A. 08h 26m, DEC. -59º03'
MAGNITUDE: c.10.0 (var)
DISTANCE 1,470 light years

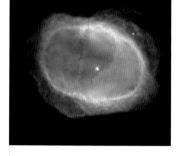

Eight-Burst Nebula NGC 3132

This bright planetary nebula lies on Vela's northern boundary, northwest of the magnitude-3.8 triple star p Velorum. Seen through small telescopes, it forms a disc of light somewhat larger than Jupiter, with a magnitude-10.0 star sitting neatly at its centre. However, appearances can be deceptive – the bright star has a binary companion of magnitude 16.0 (beyond the reach of most amateur telescopes), and it is this star that is actually responsible for the surrounding nebula – the brighter star is merely an innocent bystander.

R.A. 10h 07m, DEC. -40º26'
MAGNITUDE 9.9
DISTANCE 2,000 light years

Vela Pulsar PSR B0833-45

From the centre of the Vela Supernova Remnant, a rapidly spinning neutron star sends a beam of radio waves, X-rays and gamma rays towards Earth every 89 milliseconds. This 'pulsar' signal is created by a powerful magnetic field around the stellar corpse, which channels radiation into two narrow beams that sweep around the sky. This X-ray view of the pulsar reveals that it has a second pair of beams – jets emitted along the pulsar's axis of rotation, which spray energized particles into the surrounding nebula rather like an out-of-control firehose.

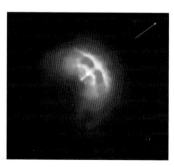

R.A. 08h 35m, DEC. -45º11'
MAGNITUDE 23.6
DISTANCE 815 light years

Pencil Nebula
NGC 2736

The brightest part of the Vela Supernova Remnant at visible wavelengths, the Pencil Nebula marks an expanding shockwave travelling through space at roughly 650,000 km per hour (400,000 mph). Where particles in the shock front collide with those in the interstellar medium, they are excited, creating a glowing wall three-quarters of a light year long, moving across space with awesome speed. The Pencil was discovered by British astronomer John Herschel, working at the Cape of Good Hope in 1835, but was not linked to the larger supernova remnant until much more recently.

R.A. 09h 00m, DEC. -45°54'
MAGNITUDE 12.0
DISTANCE 815 light years

Carina

AT A GLANCE

NAME Carina
MEANING The keel of Argo
ABBREVIATION Car
GENITIVE Carinae
R.A. 08h 42m
DEC. −63° 13'
AREA 494 (34)
BRIGHTEST STAR Canopus (α)

The keel is the brightest part of the great mythological ship Argo, sailed by Jason and the Argonauts in their quest for the golden fleece. It is unmistakable thanks to the presence of Canopus, the second-brightest star in the entire sky. It is also home to some of the richest regions of the southern Milky Way.

Canopus gets its name from a famous helmsman – the man who steered King Menelaus's fleet in its journey to Troy. It shines at magnitude -0.7, compared to Sirius's -1.4, but is a far more formidable star on any other measure: Canopus lies 30 times further away, at a distance of 315 light years, and is a yellow-white supergiant 15,000 times more luminous than the Sun and 600 times more luminous than Sirius.

The constellation's other undoubted highlight is the Carina Nebula NGC 3372. Visible to the naked eye as a bright patch in the Milky Way roughly four times the size of the Full Moon, it is a delight to explore with binoculars or a telescope, containing a wealth of interesting features (*see* pages 184–5). Away from the nebula, two other naked-eye star clusters stand out in Carina – the Wishing Well NGC 3532 and the 'Southern Pleiades' IC 2602.

Mystic Mountains
This bizarre and tortured landscape is formed by a variety of star-forming processes at work in the Carina Nebula, some 7,500 light years from Earth. The dark outcrops are similar to the 'pillars of creation' found in other nebulae, and are slowly being eroded by radiation from nearby brilliant stars.

Carina Inside View

Massive Cluster
NGC 3603

When British astronomer John Herschel discovered this cluster while working in South Africa in 1834, he at first thought that it might be globular in nature. In reality, however, NGC 3603 is a particularly dense open cluster, perhaps just a million years old and containing the greatest concentration of heavyweight stars known in the Milky Way galaxy. Though the cluster is surrounded by a dense cloud of gas and dust, fierce radiation and strong stellar winds from its stars have cleared out its surroundings to leave it beautifully framed in empty sky.

R.A. 11h 15m, DEC. -61º16'
MAGNITUDE 9.1
DISTANCE 20,000 light years

Bullet Cluster
1E-0657-558

In an empty corner of western Carina, powerful telescopes reveal this distant cluster, about 3.7 billion light years from Earth. The Bullet formed during a dramatic collision between galaxy clusters that began around 150 million years ago. This image shows the distribution of mass (blue) and X-ray emitting gas (pink) between the colliding clusters – it reveals that while the clouds of 'cluster gas' have collided head-on, both the galaxies and the mysterious 'dark matter', which provides much of the cluster's mass, have passed through each other almost unaffected.

R.A. 06h 59m, DEC. -55$^{\underline{0}}$57'
MAGNITUDE 14.2 and fainter
DISTANCE 3.7 billion light years

Star on the Brink
Eta (η) Carinae

A wide view of the Carina Nebula reveals the bulbous 'Homunculus Nebula' around Eta Carinae at its heart. One of the most remarkable stars in the sky, Eta currently shines at an unspectacular magnitude 4.6, but is an unpredictable variable: during an outburst that peaked around 1843, it briefly became the second-brightest star in the entire sky. Eta consists of a pair of blue supergiants, each with the mass of around 60 to 80 Suns and pumping out hundreds of thousands of times more energy. Both stars will end their lives as supernovae in the astronomically near future.

R.A. 10h 45m, DEC. -59$^{\underline{0}}$41'
MAGNITUDE c.4.6 (var)
DISTANCE 7,500 light years

Carina The Carina Nebula NGC 3372

Herschel's View

This vivid portrait of the Carina Nebula is a product of the European Space Agency's Herschel Space Observatory, an orbiting telescope designed for investigating the far reaches of the infrared spectrum and some of the coldest objects in the Universe. It reveals that the visible parts of the nebula are surrounded by enormous and previously unsuspected reserves of invisible gas, giving the entire nebula an estimated mass of 650,000–900,000 Suns. Despite being invisible to other instruments, this gas is clearly not immune to the forces that shape the rest of the nebula – there are many places where it has been carved into pillars or hollowed out in bubbles by the influence of the superhot stars being created within it.

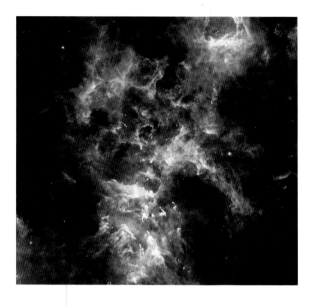

R.A. 10h 45m, DEC. -59º52'
MAGNITUDE 1.0
DISTANCE 7,500 light years

Hubble Fantasy

This segment from a mosaic constructed using 48 separate Hubble Space Telescope images covers an area of the Carina Nebula roughly a dozen light years wide – the entire nebula is more than 200 light years across. Opaque loops of dusty gas on the left of the image form part of the top of the Keyhole Nebula, a dark nebula that is one of the complex's most prominent features in amateur telescopes. In the centre, meanwhile, are several isolated 'Bok globules'. These dense pockets of gas and dust, set adrift from their origin in 'pillars of creation' may each contain an incubating star or star system.

Infrared Landscape

This beautiful near-infrared view of the Carina Nebula was captured by the HAWK-I camera on the European Southern Observatory's Very Large Telescope, which is in the Atacama Desert of northern Chile. It reveals hitherto unsuspected arcs of gas close to Eta Carinae (lower left), dark star-forming regions, previously undiscovered cooler stars shown in yellow and a spectacular central star cluster called Trumpler 14. This cluster is home to the most recent stars produced by the nebula, at around half a million years old, and may contain as many as 2,000 stars in a region just six light years across.

Crux

AT A GLANCE

NAME Crux
MEANING The southern cross
ABBREVIATION Cru
GENITIVE Crucis
R.A. 12h 27m
DEC. −60° 11'
AREA 68 (88)
BRIGHTEST STAR Acrux (α)

The smallest constellation in the sky is also one of the most distinctive, a compact group of four bright stars nestling beneath the body of the constellation Centaurus. Though known to the ancient Greeks, it was only rediscovered by European navigators in the 16th century.

The origins of this renowned constellation are surprisingly uncertain, but Crux appears to have been invented in the early 16th century. Earlier astronomers knew of its stars, but considered them to be part of Centaurus.

Three of the key stars (Alpha, Beta and Delta) are blue-white giants at 320–350 light years from Earth. Like many other bright stars in this part of the sky, they belong to the Scorpius-Centaurus OB Association and are around 10–20 million years old. Alpha Crucis (Acrux) is a bright binary pair, whose members, shining at magnitudes 1.3 and 1.7, are easily divisible with the smallest telescope.

The fourth star of the Cross, Gamma Crucis, is a red giant of magnitude 1.6, just 88 light years away. While bright Milky Way starfields form the backdrop to much of the constellation, a prominent dark nebula, known as the Coalsack, occupies its southeast quadrant.

The Jewel Box

Visible to the naked eye as a fuzzy star of magnitude 4.2, the open cluster NGC 4755 is a highlight of southern skies. It lies just to the east of Beta Crucis, with Kappa Crucis as its brightest individual star at magnitude 5.9. Binoculars or a telescope on low magnification reveal a spectacular cloud of several dozen stars, mostly blue-white but with a brilliant red giant of magnitude 7.6 forming a pleasing contrast. The Jewel Box lies far beyond the brighter stars of Crux, around 7,600 light years away, and is thought to be just 7 million years old.

Musca

AT A GLANCE

NAME Musca
MEANING The fly
ABBREVIATION Mus
GENITIVE Muscae
R.A. 12h 35m
DEC. −70° 10'
AREA 138 (77)
BRIGHTEST STAR Alpha (α)

This small, bright but somewhat jumbled constellation is easily identified thanks to its position in the Milky Way to the south of Crux. Unsurprisingly, given this location, it contains several interesting objects. Uniquely, it is a constellation that has changed its identity in quite recent times, from one insect to another.

Musca was invented in the late 16th century by Dutch astronomer Petrus Plancius, based on reports of the southern sky brought back by Dutch navigators. Originally known as Apis, the bee, it was reimagined as a fly by French astronomer Nicolas Louis de Lacaille in 1752 – in part to avoid confusion with Apus (*see* page 214).

Alpha Muscae is a massive blue-white star of magnitude 2.7, around 315 light years from Earth. The slightly more distant Beta Muscae, at magnitude 3.0, is a binary whose evenly matched components can be split with a small telescope and high magnification. Theta Muscae is another binary, with components of magnitudes 5.7 and 7.3. Both stars are hot and blue stellar heavyweights – the fainter component is a rare Wolf-Rayet star – one that has blown away significant amounts of material in stellar winds during its lifetime, revealing even hotter inner layers.

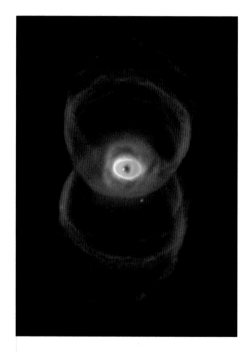

The Hourglass Nebula
Also known as MyCn 18, this planetary nebula in Musca's northwestern corner lies around 8,000 light years from Earth. With an apparent magnitude of 13.0, it is unfortunately beyond most amateur instruments, and its true structure was not appreciated until 1996, when the Hubble Space Telescope captured this remarkable image.

Circinus and Triangulum Australe

AT A GLANCE

NAME Circinus/Triangulum Australe
MEANING The compasses/The
southern triangle
ABBREVIATION Cir/TrA
GENITIVE Circini/Trianguli Australis
R.A. 14h 35m/16h 05m
DEC. −63° 02'/−65° 23'
AREA 93 (85)/110 (83)
BRIGHTEST STAR Alpha (α)/Atria (α)

Two celestial triangles – one long and thin, the other broad and almost equilateral – lie to the east of brilliant Alpha and Beta Centauri in the deep southern skies. Circinus is the Latin term for a pair of compasses.

The broad southern triangle is the older of these two constellations – first recorded in Johann Bayer's *Uranometria* star atlas of 1603, and probably invented by Dutch navigator Pieter Dirkszoon Keyser a few years before. Dating to the 1750s, the narrower triangle of the compasses is yet another invention of the French astronomer Nicolas Louis de Lacaille.

Alpha Trianguli Australis is an orange giant of magnitude 1.9, some 415 light years from Earth, while Alpha Circini is a white star of magnitude 3.2, one of a class of 'Delta Scuti' variables that changes its brightness almost imperceptibly with a period of just a few minutes. Gamma Circini is an attractive double with a blue-white component of magnitude 5.1 and a companion of magnitude 5.5 that sometimes appears yellow in contrast, but is in fact a pure white. NGC 6025 in Triangulum Australe is a tight open star cluster of magnitude 5.1, resolvable through binoculars.

Circinus Galaxy

Despite its relative closeness to Earth at just 13 million light years, the dense Milky Way star clouds that run behind Circinus kept this curious galaxy hidden and undiscovered until the 1970s. Now catalogued as ESO 97-G13, it is the closest example of an active galaxy with a bright nucleus that pours out large amounts of radiation from the region around its central supermassive black hole.

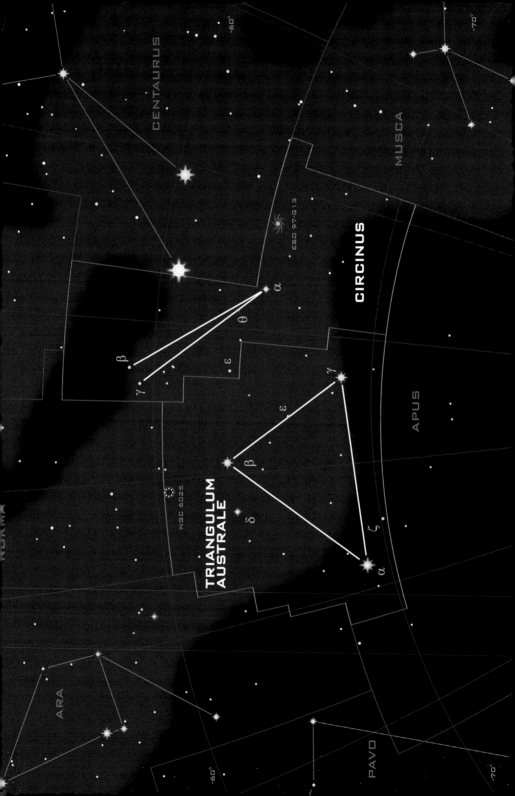

Norma and Ara

These two constellations are embedded in a rich stream of the Milky Way to the south of Scorpius. Their stars are far from impressive, but they are rich in interesting deep-sky objects such as star clusters. Despite its apparent obscurity, Ara is one of the oldest constellations in astronomical history.

AT A GLANCE

NAME Norma/Ara
MEANING The set square/The altar
ABBREVIATION Nor/Ara
GENITIVE Normae/Arae
R.A. 15h 54m/17h 22m
DEC. −51° 21'/−56° 35'
AREA 165 (74)/237 (63)
BRIGHTEST STAR Gamma-2 (γ2)/Beta (β)

Ara was recognized as a sacrificial altar by the earliest Mesopotamian stargazers. Over a thousand years later, the Greeks saw it as the altar of the Gods of Olympus and it was one of the 48 constellations described by the 2nd century astronomer Ptolemy. Norma, in contrast, was invented by Nicolas de Lacaille in the 1750s.

Gamma Normae is a line-of-sight double consisting of a magnitude-5.0 yellow supergiant some 1,450 light years from Earth, and a magnitude-4.0 yellow giant just 127 light years away. The pair can usually be separated with the naked eye. S Normae, at the heart of the NGC 6087 open star cluster, is another yellow supergiant and a variable star similar to Delta Cephei (*see* page 22), with a 9.8-day cycle that sees its magnitude vary between 6.1 and 6.8. NGC 6397 in Ara, meanwhile, is one of the nearest globular clusters at 7,200 light years, and it is easily spotted through binoculars.

Great Attractor

Hidden behind the Milky Way in Norma lies a dense galaxy cluster catalogued as Abell 3627 but generally known as the Norma Cluster. It is relatively close to Earth at 220 million light years, but best observed using X-rays (as shown here) to pierce the intervening star clouds. The cluster's location coincides with a mysterious object known as the Great Attractor – a dense concentration of mass that draws everything else in the local Universe towards it.

Corona Australis and Telescopium

AT A GLANCE

NAME Corona Australis/Telescopium
MEANING The southern crown/The telescope
ABBREVIATION CrA/Tel
GENITIVE Coronae Australis/Telescopii
R.A. 18h 39m/19h 20m
DEC. −41° 09'/−51° 02'
AREA 128 (80)/252 (57)
BRIGHTEST STAR Alfecca Meridiana (α)/
Alpha (α)

Directly to the south of Sagittarius lies an easily identified arc of stars forming the Southern Crown, a counterpart to the northern Corona Borealis. South of this lies the far more obscure constellation of the Telescope.

The Southern Crown is an ancient constellation, rising just high enough to be visible from Greek latitudes. It is often associated with Bacchus, the god of wine, or sometimes with the nearby centaur, Sagittarius. Telescopium is one of Nicolas Louis de Lacaille's 18th-century additions to the sky, and possibly the most obscure constellation of all.

Gamma Coronae Australis is an attractive binary that can be separated with small telescopes to reveal near-twin yellow stars of magnitudes 4.8 and 5.1. Eta and Kappa Coronae Australis, meanwhile, are both line-of-sight doubles, easily separated with binoculars. For amateur astronomers, the constellation's deep-sky highlight is the dense and bright globular cluster NGC 6541, 23,000 light years away and shining at magnitude 6.3. It was discovered by Nicolò Cacciatore at the Palermo Astronomical Observatory, Italy, on 19 March 1826.

R Coronae Australis
The region around the young and unpredictable variable star R Coronae Australis, 420 light years away, is home to a complex of dark, reflection and emission nebulae. As one of the nearest star-forming regions to Earth, it offers a rare glimpse of the environment in which less massive, Sun-like stars are born. Such stars do not release the high-energy radiation that illuminates more spectacular star-forming regions, so the nebula's colours are far more restrained.

Pavo

AT A GLANCE

NAME Pavo
MEANING The peacock
ABBREVIATION Pav
GENITIVE Pavonis
R.A. 19h 37m
DEC. −65° 47'
AREA 378 (44)
BRIGHTEST STAR Peacock (α)

South of Corona Australis and the obscure Telescopium, the constellation of the Peacock (*Pavo*) is one of several 'southern birds' added to the sky by Petrus Plancius in 1597, based on the reports of Dutch navigators. It is best located through the presence of the bright star, itself known as Peacock.

While most star names can be traced back to ancient Greek, Latin or Arabic roots, the origins of Alpha Pavonis's proper name are rather more recent – it was invented when the star was chose for inclusion in navigation manuals by Britain's Royal Air Force in the 1930s. The RAF insisted that all of the stars included in the manual must have proper names, so new names were invented. Accordingly Alpha Pavonis was named 'Peacock'.

At magnitude 1.9, bright-blue Peacock is a spectroscopic binary – a star that we know is a pair thanks to the analysis of its light, but which cannot be visually separated. Kappa Pavonis, meanwhile, is a pulsating Cepheid variable – a yellow supergiant 490 light years away that fluctuates between magnitudes 3.9 and 4.8 in a 9.1-day cycle. Delta Pavonis, 20 light years away and shining at magnitude 3.6, is a star of similar mass to the Sun, but somewhat older and already in the process of evolving into a red giant.

Bright Globular

NGC 6752 in northern Pavo is the third-brightest globular cluster in the entire sky, easily visible with binoculars at magnitude 5.4, and about two-thirds the diameter of the Full Moon. It lies 13,000 light years from Earth and contains more than 100,000 stars in a region of space roughly 100 light years across. This Hubble Space Telescope image covers the central ten light years of the cluster, showing that the density of stars increases exponentially towards the cluster's core.

Grus and Phoenix

Two birds, one real and one mythological, straddle the line between the bright stars Fomalhaut and Achernar in far southern skies. Both are fairly easy to identify – the mythical Phoenix from its outstretched wings as it rises from the ashes of its forebear, and the Crane from its elongated neck.

AT A GLANCE

NAME Grus/Phoenix
MEANING The crane/The phoenix
ABBREVIATION Gru/Phe
GENITIVE Gruis/Phoenicis
R.A. 22h 27m/00h 56m
DEC. −46° 21'/−48° 35'
AREA 366 (45)/469 (37)
BRIGHTEST STAR Al Nair (α)/Ankaa (α)

These 'southern birds' were first recorded by Johann Bayer in his *Uranometria* star atlas of 1603, but, like those attributed to Petrus Plancius, they started out as inventions of Pieter Dirkszoon Keyser and Frederick de Houtman, a pair of Dutch navigators who kept detailed astronomical records while trading in the East Indies during the 1590s.

Zeta Phoenicis is a triple star system: it consists of an eclipsing binary that normally shines at magnitude 3.9 but drops by half a magnitude during eclipses every 1.67 days, and a faint, close companion of magnitude 6.9. Beta Gruis, meanwhile, is a pulsating red giant that varies irregularly between magnitudes 2.0 and 2.3. Unfortunately, these constellations have no significant deep-sky objects that are suitable for observation by smaller telescopes.

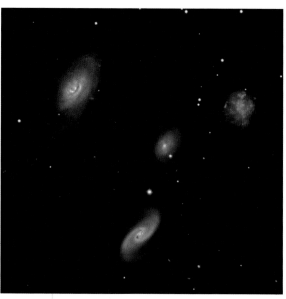

Robert's Quartet
Although at magnitudes around 14.0 these galaxies are too faint to be seen through most amateur instruments, this attractive quartet in Phoenix is still one of the sky's more famous compact galaxy groups. With an ungainly catalogue number of AM 0018-485, these four galaxies lie around 160 million light years from Earth. They are contained in an area of sky with one-tenth the diameter of the Full Moon, corresponding to a region around 75,000 light years across.

Tucana and Indus

AT A GLANCE

NAME Tucana/Indus
MEANING The toucan/The Indian
ABBREVIATION Tuc/Ind
GENITIVE Tucanae/Indi
R.A. 23h 47m/21h 58m
DEC. −65° 50'/−59° 42'
AREA 295 (48)/294 (49)
BRIGHTEST STAR Alpha (α)/
The Persian (α)

While both of these constellations are fairly faint, Tucana is home to some stunning deep-sky objects including the Small Magellanic Cloud. It can be found easily by looking southwest of bright Achernar in the constellation Eridanus. Shapeless Indus lies further to the west, but has little to recommend it.

Both of these constellations (Tucana is the fourth of the 'southern birds') were introduced to European astronomers through the reports of Dutch navigators Pieter Dirkszoon Keyser and Frederick de Houtman, who travelled widely in southeast Asia during the 1590s. There is some evidence that they may be based on earlier star patterns already recognized in the region around the East Indies.

Alpha Indi is an orange giant of magnitude 3.1, 100 light years from Earth. Medium-sized telescopes reveal two red-dwarf companions of magnitudes 11.9 and 12.5. Epsilon Indi is a nearby star, just 11.8 light years away and shining at magnitude 4.7. It is a little less massive and less luminous than the Sun, and has its own pair of faint companions – failed stars or 'brown dwarfs' that were only discovered in 2003.

Overlooked Globular

Tucana is home to two famous wonders of the heavens – the Small Magellanic Cloud and the bright globular cluster 47 Tucanae (NGC 104). As a result, another globular within Tucana's boundaries tends to be overlooked. At 30,000 light years from Earth, NGC 362 is far more distant than 47 Tucanae, and shines at magnitude 6.4, making it an easy target for night-sky observers with binoculars or a small telescope. Larger instruments will start to resolve the cluster's individual stars.

Tucana 47 Tucanae NGC 104

Celestial Glitterball

Easily visible to the naked eye at magnitude 4.9, globular cluster 47 Tucanae looks like a slightly fuzzy star on the edge of the Small Magellanic Cloud. It was French astronomer Nicolas Louis de Lacaille who first noted its non-stellar appearance in 1751, but rather like Omega Centauri's Greek 'Bayer letter', the 'Bode number' designating the object as a star has stuck. Binoculars will transform the cluster into a ball of light roughly the size of the Full Moon, but this cluster is so densely packed that even its outer edges can only be resolved with a medium-sized telescope.

R.A. ooh 24m, DEC. -72⁰05'
MAGNITUDE 4.9
DISTANCE 16,700 light years

Planet Survey

This Hubble Space Telescope view captures a field of 35,000 stars within 47 Tucanae, studied in 1999 in an effort to search for extrasolar planets in this crowded environment. The telescope stared at these stars for eight days, looking for telltale dips in their brightness caused by planetary transits (in which a planet crosses the face of its star and blocks out some of its light). Although such events are rare, this 'transit technique' has been used successfully elsewhere, and, statistically, astronomers expected to detect around 17 transits. The fact that the survey found none suggests that conditions inside globular clusters either prevent the birth of planets, or strip them away from their parent stars after formation.

Central Regions

This image from the European Southern Observatory's Very Large Telescope zooms in on the central portion of 47 Tucanae, resolving detail almost all the way to the core. At around 16,700 light years away, the cluster is one of the closer globulars to Earth, but it is also a genuinely impressive object, with a total mass equivalent to a million Suns packed into a region 120 light years across. In the centre of the cluster, the closely packed stars are separated by light days rather than light years.

Stars in Motion

This extreme close-up of 47 Tucanae's core is just one in a series of Hubble Space Telescope images that astronomers used to track the motion of more than 15,000 stars within the heart of the cluster. By comparing the speeds of the stars with their masses (which were calculated from their brightness and evolutionary state), they were able to map the complex dynamics at work inside clusters for the first time. Frequent close encounters within 47 Tucanae exchange momentum between stars, ultimately leading heavier stars to slow down in their orbits and 'sink' towards the core of the cluster, while accelerating lightweight stars and boosting their orbits towards the cluster's outer limits.

Tucana The Small Magellanic Cloud

Satellite Galaxy

Clearly visible to the naked eye in southeastern Tucana, the Small Magellanic Cloud (SMC) resembles a small, isolated fragment of the Milky Way, around seven times the diameter of the Full Moon. Binoculars and small telescopes will start to reveal a wealth of detail including star clusters and nebulae, while the constellation presents almost limitless opportunities for larger instruments. Compared to its 'big brother', the Large Magellanic Cloud, the SMC is both genuinely smaller (at around 10,000 light years across) and further away at 210,000 light years. The clusters take their name from Portuguese explorer Ferdinand Magellan, who was the first European to report them during his round-the-world voyage of 1519–21.

R.A. 00h 53m, DEC. -72º50'
MAGNITUDE 2.3
DISTANCE 210,000 light years

Stellar Nursery
N66 and NGC 346

Both the Large and Small Magellanic Clouds are classed as irregular galaxies, rich in gas and dust that supply the raw materials for ongoing star formation. N66 is one of many beautiful starbirth nebulae in the SMC, a glowing cocoon illuminated from within by the intense radiation of a recently formed star cluster known as NGC 346. At magnitude 10.3, the cluster can be located through even a small telescope, although long photographic exposures are needed to reveal the surrounding nebulosity. NGC 346 contains many of the SMC's most massive stars – the shockfront generated by their powerful stellar winds is clearly visible at the top and left of this Hubble Space Telescope image.

R.A. 00h 59m,
DEC. -72º10'
MAGNITUDE 10.3
DISTANCE 210,000
light years

Hidden Treasure
N90 and NGC 602

This stunning Hubble image showcases one of the SMC's most beautiful star-forming regions, known as N90. This cavern-like nebula is shaped by radiation and stellar winds blasting out from NGC 602, the newborn cluster of brilliant stars at its heart. As the walls of the nebula fight their losing battle against the forces of erosion, denser outcrops, similar to the 'pillars of creation' found in the Eagle Nebula (see page 102) stand out like celestial stalagmites and stalactites. Near the bottom of the picture, a distant cluster of galaxies, millions of light years beyond N90, emerges through the hazy gas.

R.A. 01h 29m, DEC. -73º34'
MAGNITUDE 13.1
DISTANCE 210,000 light years

Horologium and Reticulum

AT A GLANCE

NAME Horologium/Reticulum
MEANING The clock/The reticle
ABBREVIATION Hor/Ret
GENITIVE Horologii/Reticuli
R.A. 03h 17m/03h 55m
DEC. −53° 20'/−59° 60'
AREA 249 (58)/114 (82)
BRIGHTEST STAR Alpha (α)/Alpha (α)

Tucked beneath the southern reach of the celestial river Eridanus, these two constellations are faint and obscure. The Clock (intended to represent the form of a swinging clock pendulum) is particularly hard to spot, but the tight diamond of the Reticle or Crosshair is at least a little clearer.

Both of these rather threadbare constellations stand mainly as a testimony to the imagination and determination of their inventor, French astronomer Nicolas Louis de Lacaille, in his 18th-century effort to fill every last gap in the southern sky.

Alpha Horologii, marking the pivot at the top of the clock's pendulum, is an orange giant of magnitude 3.9, around 117 light years from Earth. Alpha Reticuli is a binary consisting of a yellow giant of magnitude 3.4, and a faint red dwarf of magnitude 12.0, visible with a medium-sized telescope. Perhaps the most famous star in either constellation, however, is Zeta Reticuli, a binary pair of Sun-like stars 39 light years away, shining at magnitudes 5.5 and 5.2 and divisible with binoculars. Zeta is well known as the supposed home of the 'grey' aliens that UFO enthusiasts believe abducted an American couple, Betty and Barney Hill, in 1961.

NGC 1559

This barred spiral galaxy in Reticulum is unfortunately beyond the reach of most amateur telescopes, but views from instruments such as the Hubble Space Telescope reveal it to be a beautiful and active galaxy with a bright core and well-defined spiral arms alive with star formation.

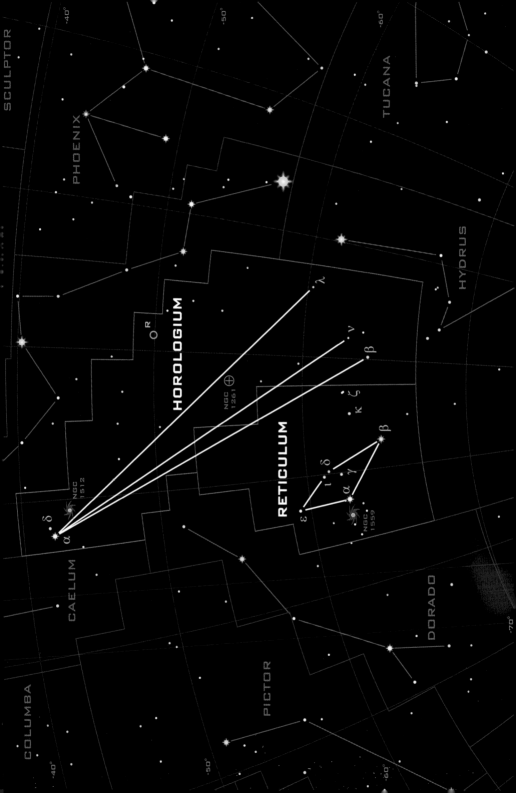

Pictor and Dorado

These two faint and indistinct constellations lie south of a line joining the bright stars Canopus in Carina and Achernar in Eridanus. While Dorado is home to the spectacular Large Magellanic Cloud, a satellite galaxy of the Milky Way, Pictor has no such impressive attractions.

AT A GLANCE

NAME Pictor/Dorado
MEANING The easel/The dorado
ABBREVIATION Pic/Dor
GENITIVE Pictoris/Doradus
R.A. 05h 42m/05h 15m
DEC. −53° 28'/−59° 23'
AREA 247 (59)/179 (72)
BRIGHTEST STAR Alpha (α)/Alpha (α)

Dorado is another invention of the Dutch navigators Keyser and de Houtman, added to the sky around 1600. Although its name translates as 'goldfish', it was probably intended to represent the Hawaiian *mahi-mahi* fish. Pictor, meanwhile, is another of Nicolas Louis de Lacaille's southern constellations from the 1750s.

Beta Doradus is one of the brightest variable stars in the sky – a pulsating yellow supergiant that varies in brightness between magnitudes 3.5 and 4.1 in a 9.9-day cycle. This Cepheid variable lies 1,040 light years from Earth, and its variations can easily be tracked with reference to other nearby stars. Delta Pictoris, meanwhile, is an eclipsing binary that dips from magnitude 4.7 to 4.9 every 40 hours as one of its hot blue-white stars passes in front of the other. Its variations are just detectable to the human eye.

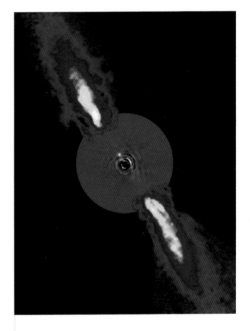

Beta Pictoris
Pictor's most famous object is an unassuming white star of magnitude 3.9, 63 light years from Earth. It owes its fame to a broad disc of gas and dust that swirls around it, extending to more than 40 times the width of Neptune's orbit. This 'protoplanetary disc' was the first object of its kind to be discovered (thanks to the excessive infrared radiation it emits, as seen in the image above).

Dorado The Large Magellanic Cloud

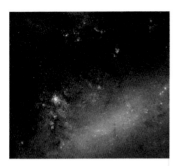

Irregular Satellite
The Large Magellanic Cloud

(LMC) is the largest of the Milky Way's satellite galaxies, following a 1.5 billion-year orbit shared with the Small Magellanic Cloud (*see* page 204). With a diameter of 20,000 light years, it resembles an isolated patch of our own galaxy, 20 times the width of the Full Moon, on the border of Dorado and Mensa. The LMC is easily spotted with the naked eye, and binoculars reveal a broad bar of stars that is its major feature – although usually classed as irregular, the LMC shows some structure and is sometimes called a 'one-armed spiral'. Telescopes of any size are ideal for exploring its population of nebulae and star clusters.

R.A. 05h 24m, DEC. -69º45'
MAGNITUDE 0.1
DISTANCE 179,000 light years

Starry Vista
LH 95

Although dwarfed by the nearby Tarantula Nebula, this star-forming region of the LMC has provided a wealth of information about the way stars form inside this gas- and dust-rich galaxy. Until recently, the nebula was known only from bright and hot blue-white stars, with up to three times the mass of the Sun. In 2006, however, a study using the Hubble Space Telescope revealed more than 2,500 infant stars that have not yet settled down onto the 'main sequence' of stellar evolution (the yellow and orange stars in the image below right). These include red dwarfs with as little as one-third the mass of the Sun.

R.A. 05h 37m,
DEC. -66º22'
MAGNITUDE 11.1
DISTANCE 180,000
light years

Tarantula Nebula
NGC 2070

The Tarantula Nebula is one of the largest star-forming regions in our Local Group of galaxies. Binoculars or a small telescope will show it as a collection of long gaseous tendrils resembling the legs of a gigantic spider. Overall, the complex is around 1,000 light years across – if it was transplanted to the current location of the Orion Nebula M42, it would cover some 30 degrees of the sky and be bright enough to cast shadows. The nebula is thought to owe its size and intensity to its position on the leading edge of the LMC, where it suffers from compression effects as the galaxy moves around its orbit.

R.A. 05h 39m,
DEC. -69º06'
MAGNITUDE 8.0
DISTANCE 180,000
light years

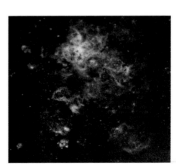

Superstars
R136

Small telescopes are enough to spot the star cluster at the heart of the Tarantula Nebula, but they cannot hint at its awesome scale. This dense ball of heavyweight blue stars is just 1–2 million years old and pumps out intense ultraviolet radiation that excites gas throughout the enormous nebula. At its heart sits a tight knot of stars called R136a, recently separated into individual components that include R136a1, the most massive star known up to this time. This stellar monster has 265 times the mass of the Sun and 10 million times its luminosity – it is about as big as a star can get before it begins tearing itself apart.

R.A. 05h 39m, DEC. -69º06'
MAGNITUDE 9.5
DISTANCE 180,000 light years

Mensa and Volans

AT A GLANCE

NAME Mensa/Volans
MEANING Table Mountain/
 The flying fish
ABBREVIATION Men/Vol
GENITIVE Mensae/Volantis
R.A. 05h 25m/07h 48m
DEC. −77° 30'/−69° 48'
AREA 153 (75)/141 (76)
BRIGHTEST STAR Alpha (α)/Gamma (γ)

These two constellations of the far south are lacking both in bright stars and noticeable patterns, but are still relatively easy to locate since Volans lies directly to the south of the bright constellation Carina, and Mensa lies between the Large Magellanic Cloud in Dorado and the south celestial pole.

Volans was one of the new southern constellations invented by the Dutch navigators Keyser and de Houtman in the 1590s, while Mensa was added to the sky by Nicolas Louis de Lacaille in the 1750s, inspired by the daily sight of Table Mountain in South Africa with its top shrouded in cloud. Lacaille had conducted observations of the southern sky from the slopes of this mountain.

Alpha Mensae is an unimpressive yellow dwarf with 80 per cent of the Sun's luminosity, only visible to the naked eye, at magnitude 5.1, because it lies just 33 light years away. Volans at least has two attractive binaries to offer, both divisible with a small telescope. Epsilon Volantis is a blue-white star of magnitude 4.4 with a yellow companion of magnitude 8.1, while Gamma Volantis pairs an orange giant of magnitude 3.8 with a white star of magnitude 5.7.

The Meathook Galaxy
At magnitude 11.2, this beautiful and unusual spiral is only visible through larger amateur telescopes, but it is nevertheless a highlight of far southern skies. Officially catalogued as NGC 2442, the galaxy's remarkable S-shape is formed by two asymmetric spiral arms, one of which is unusually compressed, while the other is elongated. The distortion is thought to be the result of a recent close encounter with another nearby galaxy, some 50 million light years from Earth.

Chamaeleon and Apus

AT A GLANCE

NAME Chamaeleon/Apus
MEANING The chameleon/
    The bird of paradise
ABBREVIATION Cha/Aps
GENITIVE Chamaeleontis/Apodis
R.A. 10h 42m/16h 09m
DEC. −79° 12'/−75° 18'
AREA 132 (79)/206 (67)
BRIGHTEST STAR Alpha (α)/Alpha (α)

These two far southern constellations are faint and shapeless, but easily located between the south celestial pole and the brighter stars of Carina, Musca and Triangulum Australe. Both these constellations are inventions of the Dutch navigators Pieter Dirkszoon Keyser and Frederick de Houtman.

Keyser and de Houtman based many of their new star patterns on creatures they encountered on their trading mission to the East Indies. They were introduced to European astronomers through a celestial globe produced by Petrus Plancius in 1598. They were first depicted in Johann Bayer's star atlas *Uranometria* in 1603. Chamaeleon is sometimes also referred to as the Frying Pan in Australia.

While this area of the sky is devoid of deep-sky objects suitable for amateurs, the two constellations offer a number of double stars. Delta Chamaeleontis is a line-of-sight alignment consisting of stars at magnitudes 4.4 and 5.5, 355 and 365 light years from Earth respectively. Delta Apodis is a pairing of red and orange giants at magnitudes 4.7 and 5.3 – both are around 700 light years away, and they may be a true binary. Epsilon Chamaeleontis is a binary pair of twin white stars, magnitudes 5.4 and 6.0, that can only be split with a medium-sized telescope.

Chamaeleon Complex
The far southern skies contain a broad swathe of dark clouds in the constellation of Chamaeleon. This star-forming region some 500 light years from Earth remains inconspicuous because it is giving birth to lightweight, Sun-like stars, which do not excite the surrounding nebulosity enough to make it glow.

Hydrus and Octans

AT A GLANCE

NAME Hydrus/Octans
MEANING The little water snake/
The octant
ABBREVIATION Hyi/Oct
GENITIVE Hydri/Octantis
R.A. 02h 21m/23h 00m
DEC. −69° 57'/−82° 09'
AREA 243 (61)/291 (50)
BRIGHTEST STAR Beta (β)/Nu (ν)

The region around the south celestial pole is sadly bereft of bright stars, marked by the obscure constellation of the Octant. In terms of objects of interest, the nearby Little Water Snake fares little better. Hydrus forms a faint zigzag in the sky south of the bright star Achernar in Eridanus.

Invented by the Dutch navigators Pieter Dirkszoon Keyser and Frederick de Houtman, the constellation Hydrus was popularized by Petrus Plancius around 1600. Octans, meanwhile, is one of Nicolas Louis de Lacaille's inventions from the 1750s, named after a now-obsolete navigational instrument, the octant, which was used to measure the distance between a celestial object and the horizon.

Beta Hydri is one of the brightest stars in our solar neighbourhood, 24 light years from Earth. It has 10 per cent more mass than the Sun, but is somewhat older. Having exhausted its primary fuel supply, it has begun to swell and brighten in an effort to keep shining – a process that will eventually transform it into a brilliant red giant. Lambda Octantis is an attractive double with yellow and white components of magnitudes 5.4 and 7.7. It lies 435 light years from Earth and is easily separated in a small telescope.

Polar Stars
A stunning long-exposure view from the Australian coast shows the south celestial pole hanging over the Southern Ocean, with the horizon lit by a beautiful display of the Aurora Australis or Southern Lights. The sky's southern pivot is sadly lacking in a bright star equivalent to the northern Polaris – the closest naked-eye star to the pole itself is the obscure Sigma Octantis, but at magnitude 5.4 it is only visible under dark skies.

Glossary

Active galaxy A galaxy that emits large amounts of energy from its central regions, probably generated as matter falls into a supermassive black hole at the heart of the galaxy.

Astronomical unit A unit of measurement widely used in astronomy, equivalent to Earth's average distance from the Sun – which is roughly 150 million km (93 million miles).

Barred spiral galaxy A spiral galaxy in which the arms are linked to the hub by a straight bar of stars and other material.

Bayer letters Greek letters assigned to the brighter stars in a constellation, offering a general indication of their order of brightness.

Binary star A pair of stars in orbit around one another. Because the stars in a binary pair were usually born at the same time, they allow us to compare the way that stars with different properties evolve.

Black hole A superdense point in space, usually formed by a collapsing stellar core of at least five solar masses. Its gravity is so powerful that even light cannot escape it.

Brown dwarf A so-called 'failed star' that lacks the mass to trigger fusion of hydrogen in its core, and so never starts to shine properly. Instead brown dwarfs radiate low energy radiation (mostly infrared).

Celestial equator An imaginary line around the sky lying directly above Earth's equator, and separating the sky into northern and southern hemispheres.

Celestial pole A point in the sky that happens to lie directly above one of Earth's poles of rotation.

Celestial sphere An imaginary spherical shell around the Earth that provides a useful 'model' for mapping the sky.

Circumpolar Descriptive of a constellation that lies close to either the north or south celestial pole, and as a result never sets in the skies across a large part of a particular hemisphere.

Constellation Technically, one of 88 areas of the sky defined by boundary lines in right ascension and declination. Traditionally, however, constellations are patterns in the sky produced by joining certain stars with imaginary lines.

Dark nebula A cloud of interstellar gas and dust that absorbs light, and only becomes visible when silhouetted against a field of stars or other nebulae.

Declination (Dec.) A measurement in the equatorial coordinate system, roughly analogous to latitude on Earth's surface, used in conjunction with right ascension.

Eclipsing binary A binary star system in which one star regular passes in front of the other, causing the system's combined brightness to drop periodically.

Ecliptic A path around the sky describing the apparent motion of the Sun, produced as Earth orbits around it over the course of a complete year.

Elliptical galaxy A galaxy consisting of stars in orbits that have no particular orientation, and generally lacking in star-forming gas. Ellipticals include both the smallest and the largest galaxies known, and are typically composed of long-lived, low-mass red and yellow stars.

Elliptical orbit An orbit in the form of a stretched circle, with one axis (the 'semi-major' axis) typically longer than the other.

Emission nebula A cloud of gas in space that glows at very specific wavelengths, producing a spectrum full of 'emission lines'. These nebulae are usually energized by the high-energy light of nearby stars, and are often associated with star-formation regions.

Equatorial coordinates The most widely used astronomical coordinate system, measuring the locations of celestial objects in terms of their declination relative to the celestial equator, and their right ascension relative to the First Point of Aries.

Flamsteed numbers A system of numbers assigned to the fainter naked-eye stars within constellations (those not given Bayer letters). In general, Flamsteed numbers increase from east to west across a constellation.

Galaxy An independent system of stars, gas and other material with a size measured in thousands of light years, containing anything from millions to billions of stars.

Globular cluster A dense ball of ancient and long-lived stars, typically found in orbit around a galaxy such as the Milky Way.

Infrared Electromagnetic radiation with slightly less energy than visible light. Infrared radiation is typically emitted by warm objects too cool to glow visibly.

Irregular galaxy A galaxy with no obvious structure, generally rich in gas, dust and star-forming regions.

Light year A common unit of astronomical measurement, equivalent to the distance travelled by light in one year – it is approximately 9.5 million million km (5.9 trillion miles).

Luminosity A measure of a star's energy output. Luminosity is technically measured in watts, but stars are so luminous that it is often simpler to compare them to the Sun.

Magnitude Apparent magnitude is a measure of a star's brightness as seen from Earth, using a simple number – the lower the number, the brighter the object. The brightest star in the sky, Sirius, shines at magnitude –1.4, while the faintest naked-eye stars are around magnitude 6.0.

Main sequence A term used to describe the longest phase in a star's life, during which it is relatively stable and shines by fusing hydrogen (the lightest element) into helium (the next lightest) at its core.

Multiple star A system of two or more stars in orbit around a shared 'centre of gravity' (pairs of stars are also called binaries). Most of the stars in our galaxy are members of multiple systems rather than individuals like the Sun.

Nebula Any cloud of gas or dust floating in space. Nebulae are the material from which stars are born, and into which they are scattered again at the end of their lives.

Neutron star The collapsed core of a massive star, formed in a supernova explosion. A neutron star consists of compressed subatomic particles, and is the densest known object –

although in the most massive stars, the core can collapse past this stage to form a black hole. Many neutron stars initially behave as pulsars.

Nova A binary star system prone to dramatic outbursts, in which a white dwarf is pulling material from a companion star, building up a layer of gas around itself that then burns away in a violent nuclear explosion.

Open cluster A large group of bright young stars that have recently been born from the same star-forming nebula, and may still be embedded in its gas clouds.

Planetary nebula An expanding cloud of glowing gas sloughed off from the outer layers of a dying red giant star as it transforms into a white dwarf stellar remnant.

Pole star A star that happens to lie close to one or other celestial pole, and which therefore remains more or less fixed in the sky.

Precession A gradual rotation of the celestial sphere relative to the stars, created as tidal forces from the Sun and Moon cause Earth's axis of rotation to slowly change its orientation in space.

Protostar A star that is still coalescing from the collapse of a gas cloud under its own gravity. As the centre of the cloud heats up, it may emit infrared radiation.

Pulsar A rapidly spinning neutron star (collapsed stellar core) with an intense magnetic field that channels its radiation out along two narrow beams that sweep across the sky like a cosmic lighthouse.

Red dwarf A star with considerably less mass than the Sun – small, faint and with a low surface temperature. Red dwarfs fuse hydrogen into helium in their cores very slowly and live for much longer than the Sun, despite their size.

Red giant A star passing through a phase of its life where its luminosity has increased hugely, causing its outer layers to expand and its surface to cool. Stars usually enter red giant phases when they exhaust fuel supplies in their core.

Reflection nebula A cloud of interstellar gas and dust that shines as it reflects or scatters light from nearby stars.

Right ascension (R.A.) A measurement of celestial coordinates roughly analogous to longitude on Earth's surface, used in conjunction with declination.

Spiral galaxy A galaxy consisting of a hub of old yellow stars, surrounded by a flattened disc of younger stars, gas and dust, with spiral arms marking regions of current star formation.

Supernova An enormous stellar explosion marking the death of a star far more massive than the Sun.

Supernova remnant A cloud of shredded, superhot gas that marks the aftermath of a supernova explosion. The term may also be used to describe the strange objects left behind in the wreckage by the collapsed stellar core – either a neutron star or a black hole.

Variable star A star that changes its apparent brightness over time. Variations may be small or large, and may happen over very short or very long periods. They may be caused by interacting systems (as in eclipsing binaries and novae) or by genuine pulsations in the size and luminosity of an individual star.

Zodiac A band of a dozen ancient constellations marking the Sun's annual path around the sky (known as the ecliptic).

Index

Acknowledgements

1: Gordon Garradd/Science Photo Library; 6: Rick Witacre/Shutterstock; 10: Jason Auch; 12: NASA, ESA, and the Hubble Heritage (STScI/AURA)-ESA/Hubble Collaboration; 14: Babak Tafreshi, TWAN/Science Photo Library; 18: David Parker/Science Photo Library; 20: Credit: NASA, ESA, HEIC, and The Hubble Heritage Team (STScI/AURA) Acknowledgment: R. Corradi (Isaac Newton Group of Telescopes, Spain) and Z. Tsvetanov (NASA); 22: T.A. Rector (University of Alaska Anchorage) and WIYN/NOAO/AURA/NSF; 24: Image Data - Subaru Telescope (NAOJ), Hubble Legacy Archive; Processing - Robert Gendler; 26: NASA/DOE/Fermi LAT Collaboration, CXC/SAO/JPL-Caltech/Steward/O. Krause et al., and NRAO/AUI; 28: T. A. Rector & B. A. Wolpa, NOAO, AURA, NSF; 30l: Research by Kloppenborg et al Nature 464, 870-872 (8 April 2010)". Graphic by John D. Monnier, University of Michigan.; 30r: Adam Block/Mount Lemmon SkyCenter/University of Arizona; 31: Adam Block/NOAO/AURA/NSF; 32: R. Barrena and D. López (IAC).; 34a: ESO; 34b: Robert Williams and the Hubble Deep Field Team (STScI) and NASA; 34-35r: NASA, ESA and the Hubble Heritage Team STScI/AURA). Acknowledgment: A. Zezas and J. Huchra (Harvard-Smithsonian Center for Astrophysics); 36-37l: NASA, ESA and the Hubble Heritage Team STScI/AURA). Acknowledgment: J. Gallagher (University of Wisconsin), M. Mountain (STScI) and P. Puxley (NSF].; 37a: NASA/JPL-Caltech/STScI/CXC/UofA/ESA/AURA/JHU; 37b: NASA, ESA and R. de Grijs (Inst. of Astronomy, Cambridge, UK); 38: X-ray: NASA/CXC/Univ. of Maryland/A.S. Wilson et al.; Optical: Pal.Obs. DSS; IR: NASA/JPL-Caltech; VLA: NRAO/AUI/NSF; 40a: Credit for the NICMOS Image: NASA, ESA, M. Regan and B. Whitmore (STScI), and R. Chandar (University of Toledo), Credit for the ACS Image: NASA, ESA, S. Beckwith (STScI), and the Hubble Heritage Team (STScI/AURA); 40b: H. Ford (JHU/STScI), the Faint Object Spectrograph IDT, and NASA; 40-41r: NASA, ESA, S. Beckwith (STScI), and The Hubble Heritage Team STScI/AURA); 42: ESO/L. Calçada; 44: Jack Burgess/Adam Block/NOAO/AURA/NSF; 46: ESA/Hubble and NASA; 48: The Hubble Heritage Team (AURA/STScI/NASA); 50: ESO; 52: NASA, ESA, the Hubble Heritage (STScI/AURA)-ESA/Hubble Collaboration, and the Digitized Sky Survey 2. Acknowledgment: J. Hester (Arizona State University) and Davide De Martin (ESA/Hubble); 54al: NASA/Marshall Space Flight Center; 54ar: Image courtesy of NRAO/AUI; 54b: Richard Yandrick (Cosmicimage.com); 55: T. A. Rector/University of Alaska Anchorage and WIYN/NOAO/AURA/NSF; 56a: Zachary Grillo & the ESA/ESO/NASA Photoshop FITS Liberator; 56b: T.A. Rector/University of Alaska Anchorage and NOAO/AURA/NSF; 57: NASA/JPL-Caltech/L. Rebull (SSC/Caltech); 58: Adam Block/NOAO/AURA/NSF; 60a: NASA/JPL-Caltech/UCLA; 60b: NASA, ESA and T. Lauer (NOAO/AURA/NSF); 60-61r: Adam Evans; 62: Caltech, Palomar Observatory, Digitized Sky Survey; Courtesy: Scott Kardel; 64a: N.A.Sharp/NOAO/AURA/NSF; 64b: Jean-Charles Cuillandre (CFHT) & Giovanni Anselmi (Coelum Astronomia), Hawaiian Starlight; 65: NASA, ESA, NRAO and L. Frattare (STScI). Science Credit: X-ray: NASA/CXC/IoA/A.Fabian et al.; Radio: NRAO/VLA/G. Taylor; Optical: NASA, ESA, the Hubble Heritage (STScI/AURA)-ESA/Hubble Collaboration, and A. Fabian (Institute of Astronomy, University of Cambridge, UK); 66: Gemini Observatory, GMOS Team; 68: Richard and Leslie Maynard/Adam Block/NOAO/AURA/NSF; 70a: P.Massey (Lowell), N.King (STScI), S.Holmes (Charleston), G.Jacoby (WIYN)/AURA/NSF; 70b: NASA/JPL-Caltech; 71: NASA, Hui Yang University of Illinois ODNursery of New Stars; 72: Eckhard Slawik/Science Photo Library; 74a: NASA, ESA and AURA/Caltech; 74b: NASA and The Hubble Heritage Team (STScI/AURA) Acknowledgment: George Herbig and Theodore Simon (Institute for Astronomy, University of Hawaii); 75: NASA/JPL-Caltech/J. Stauffer (SSC/Caltech); 76: Credits for Optical Image: NASA/HSTASU/J. Hester et al. Credits for X-ray Image: NASA/CXC/ASU/J. Hester et al.; r: NASA/CXC/ASU/J. Hester et al.; 77: NASA, ESA, J. Hester and A. Loll (Arizona State University); 78: NASA, Andrew Fruchter and the ERO Team (Sylvia Baggett (STScI), Richard Hook (ST-ECF), Zoltan Levay (STScI)]; 80: Celestial Image Co./Science Photo Library; 82: MASIL Imaging Team; 84a: ESO; 84cl: ESO/Oleg Maliy; 84br: ESO; 84-85r: NASA, ESA and the Hubble Heritage (STScI/AURA)-ESA/Hubble Collaboration. Acknowledgement: Davide De Martin and Robert Gendler; 86: NOAO/AURA/NSF and N.A. Sharp (NOAO); 88al: NASA/ESA/Hubble Heritage Team (STScI/AURA); 88ar: ESO; 88b: G. Fritz Benedict, Andrew Howell, Inger Jorgensen, David Chapell (University of Texas), Jeffery Kenney (Yale University), and Beverly J. Smith (CASA, University of Colorado), and NASA; 89: G. Fazio (Harvard-Smithsonian Astrophysical Observatory) L. Jenkins (Goddard Space Flight Center) A. Hornschemeier (Goddard Space Flight Center) B. Mobasher (Space Telescope Science Institute) D. Alexander (University of Durham, UK) F. Bauer (Columbia University); 90: NOAO/AURA/NSF; 92a: NASA and The Hubble Heritage Team (STScI/AURA); 92b: NASA/JPL-Caltech/R. Kennicutt (University of Arizona) and the SINGS Team; 93: NOAO/AURA/NSF; 94-95l: NASA and The Hubble Heritage Team (STScI/AURA); 95ar: NASA/JPL-Caltech and the Hubble Heritage Team (STScI/AURA); 95br: X-ray: NASA/UMass/Q.D.Wang et al.; Optical: NASA/STScI/AURA/Hubble Heritage; Infrared: NASA/JPL-Caltech/Univ. AZ/R. Kennicutt/SINGS Team; 96: Image Credit: Lynette Cook; 98: NASA/JPL-Caltech/L. Allen (Harvard-Smithsonian CfA & Gould's Belt Legacy Team; 100a: NASA and The Hubble Heritage Team (STScI/AURA); 100c: ESA/Hubble & NASA; 100b: NASA/ESA, Friendlystar; 101: NASA, J. English (U. Manitoba), S. Hunsberger, S. Zonak, J. Charlton, S. Gallagher (PSU), and L. Frattare (STScI); 102a: ESO; 102b: NASA, ESA, STScI, J. Hester and P. Scowen (Arizona State University); 103: NASA, ESA, and The Hubble Heritage Team (STScI/AURA); 104: NASA & ESA; 106: NASA, ESA, the Hubble Heritage (STScI/AURA)-ESA/Hubble Collaboration, and A. Evans (University of Virginia, Charlottesville/NRAO/Stony Brook University); 107: Adam Block/Science Photo Library; 108: NASA, The Hubble Heritage Team (STScI/AURA); 110: NASA/WikiSky; 112: NASA, ESA, and the Hubble SM4

ERO Team; 114: Bruce Balick (University of Washington), Jason Alexander (University of Washington), Arsen Hajian (U.S. Naval Observatory), Yervant Terzian (Cornell University), Mario Perinotto (University of Florence, Italy), Patrizio Patriarchi (Arcetri Observatory, Italy), NASA; 116a: NASA/JPL-Caltech/Univ. of Arizona; 116c: ESO; 116b: ESO/VISTA/J. Emerson. Acknowledgment: Cambridge Astronomical Survey Unit; 117: NASA, ESA, C.R. O'Dell (Vanderbilt University), M. Meixner and P. McCullough (STScI); 118: NASA/JPL-Caltech/C. Martin (Caltech)/M. Seibert(OCIW); 120: Nigel Sharp/NOAO/AURA; 122a: Andrea Dupree (Harvard-Smithsonian CfA), Ronald Gilliland (STScI), NASA and ESA; 122c: NASA and The Hubble Heritage Team (STScI/AURA). Acknowledgment: C. R. O'Dell (Vanderbilt University); 122b: © Stocktrek Images/Corbis; 123: ESO/J. Emerson/VISTA. Acknowledgment: Cambridge Astronomical Survey Unit; 124a: NASA, ESA and L. Ricci (ESO); 124b: NASA; K.L. Luhman (Harvard-Smithsonian Center for Astrophysics, Cambridge, Mass.); and G. Schneider, E. Young, G. Rieke, A. Cotera, H. Chen, M. Rieke, R. Thompson (Steward Observatory, University of Arizona, Tucson, Ariz.); 124-125r: NASA,ESA, M. Robberto (Space Telescope Science Institute/ESA) and the Hubble Space Telescope Orion Treasury Project Team; 126: NASA, ESA and H.E. Bond (STScI); 128a: NASA/ESA and The Hubble Heritage Team (AURA/STScI).; 128b: Nick Wright (Nwright6302); 129: ESO; 130: NASA, H.E. Bond and E. Nelan (Space Telescope Science Institute, Baltimore, Md.); M. Barstow and M. Burleigh (University of Leicester, U.K.); and J.B. Holberg (University of Arizona); 132a: NASA/JPL-Caltech/ Harvard-Smithsonian CfA; 132b: ESO/B. Bailleul; 132-133r: NASA and The Hubble Heritage Team (STScI); 133b: NASA, ESA, and R. Humphreys (University of Minnesota); 134: NASA, ESA, the Hubble Heritage (STScI/AURA)-ESA/Hubble Collaboration, and W. Keel (University of Alabama); 136al: NASA and The Hubble Heritage Team (STScI/AURA). Acknowledgment: C. Conselice (U. Wisconsin/STScI); 136ar: NASA/WikiSky; 136b: ESO; 137: NASA, ESA, and the Hubble Heritage Team (STScI/ AURA). Acknowledgment: R. O'Connell (University of Virginia) and the Wide Field Camera 3 Science Oversight Committee; 138: NASA, ESA, and the Hubble Heritage Team (STScI/AURA)-ESA/Hubble Collaboration; 140: Eckhard Slawik/Science Photo Library; 142a: ESO/WFI (Optical); MPIfR/ESO/APEX/A.Weiss et al. (Submillimetre); NASA/CXC/CfA/R.Kraft et al. (X-ray); 142b: ALMA (ESO/NAOJ/NRAO); ESO/Y. Beletsky; 143: NASA, ESA, and the Hubble Heritage (STScI/AURA)-ESA/Hubble Collaboration. Acknowledgment: R. O'Connell (University of Virginia) and the WFC3 Scientific Oversight Committee; 144a: ESO; 144b: NASA and The Hubble Heritage Team (STScI/AURA). Acknowledgment: A. Cool (SFSU); 145: NASA, ESA, and the Hubble SM4 ERO Team; 146: NASA and The Hubble Heritage Team (STScI/AURA). Acknowledgment: C.R. O'Dell (Vanderbilt University); 148: N.A.Sharp, Mark Hanna, REU program/NOAO/AURA/NSF; 150ar: NASA/Wikisky; 150cl: NASA, The Hubble Heritage Team, STScI, AURA; 150b: Robert Gendler/Science Photo Library; 151: Royal Observatory, Edinburgh/AAO/Science Photo Library; 152: ESO; 154a: ESA/Hubble & NASA; 154b: ESO/INAF-VST/OmegaCAM. Acknowledgement: OmegaCen/Astro-WISE/Kapteyn Institute; 154-155r: ESO; 156a: ESO/S. Guisard; 157a: NASA/CXC/MIT/F. Baganoff, R. Shcherbakov et al.; 156-157b: NASA/JPL-Caltech/ESA/CXC/STScI; 157: Hubble/Wikisky; 160: NASA, ESA and P. Kalas (University of California, Berkeley, USA); 162: ESO and Digitized Sky Survey 2. Acknowledgment: Davide De Martin; 164ar: ESO/R. Gendler; 164cl: ESO/J. Emerson/VISTA. Acknowledgment: Cambridge Astronomical Survey Unit; 164b: ESO; 165: NASA, ESA, S. Beckwith (STScI) and the HUDF Team; 166: NASA/JPL-Caltech/T. Pyle (SSC); 168: WikiSky; 170: NASA/JPL-Caltech/SSC; 172: NASA, ESA and Orsola De Marco (Macquarie University); 174: ESO; 176: ESO; 178ar: NASA, ESA, and P. Hartigan (Rice University); 178b: The Hubble Heritage Team (STScI/AURA/NASA); 179: ESO; 180: NASA, ESA, and M. Livio and the Hubble 20th Anniversary Team (STScI); 182al: NASA, ESA, R. O'Connell (University of Virginia), F. Paresce (National Institute for Astrophysics, Bologna, Italy), E. Young (Universities Space Research Association/Ames Research Center), the WFC3 Science Oversight Committee, and the Hubble Heritage Team (STScI/AURA); 182b: X-ray: NASA/CXC/CfA/M.Markevitch et al.; Optical: NASA/STScI; Magellan/U.Arizona/D.Clowe et al.; Lensing Map: NASA/STScI; ESO WFI; Magellan/U.Arizona/D.Clowe et al.; 182-183r: NOAO/AURA/NSF; 184a: ESA/PACS/SPIRE/Thomas Preibisch, Universitäts-Sternwarte München, Ludwig-Maximilians-Universität München, Germany.; 184b: Credit for Hubble Image: NASA, ESA, N. Smith (University of California, Berkeley), and The Hubble Heritage Team (STScI/AURA). Credit for CTIO Image: N. Smith (University of California, Berkeley) and NOAO/AURA/NSF; 185: ESO/T. Preibisch; 186: ESO/Y. Beletsky; 188: Raghvendra Sahai and John Trauger (JPL), the WFPC2 science team, andNASA/ESA); 190: NASA, Andrew S. Wilson (University of Maryland); Patrick L. Shopbell (Caltech); Chris Simpson (Subaru Telescope); Thaisa Storchi-Bergmann and F. K. B. Barbosa (UFRGS, Brazil); and Martin J. Ward (University of Leicester, U.K.); 192: X-ray: NASA/CXC/UVa/M. Sun et al; H-alpha/Optical: SOAR/MSU/NOAO/UNC/CNPq-Brazil/M.Sun et al.; 194: ESO; 196: NASA/WikiSky; 198: ESO; 200: NASA/JPL-Caltech/University of Virginia/R. Schiavon (Univ. of Virginia); 202al: ESA/Hubble (Davide De Martin), the ESA/ESO/NASA Photoshop FITS Liberator & Digitized Sky Survey 2; 202c: ESO; 202b: NASA and Ron Gilliland (Space Telescope Science Institute); 203: NASA, ESA, and G. Meylan (Ecole Polytechnique Federale de Lausanne); 204a: ESA/Hubble and Digitized Sky Survey 2. Acknowledgements: Davide De Martin (ESA/Hubble); 204b: NASA, ESA and A. Nota (ESA/STScI, STScI/AURA); 205: NASA, ESA, and the Hubble Heritage Team (STScI/ AURA) - ESA/Hubble Collaboration; 206: NASA/WikiSky; 208: ESO/A.-M. Lagrange et al.; 210al: ESO; 210cr: NASA, ESA, and the Hubble Heritage Team (STScI/AURA)-ESA/Hubble Collaboration. Acknowledgment: D. Gouliermis (Max Planck Institute for Astronomy, Heidelberg); 210b: ESO/R. Fosbury (ST-ECF); 211: NASA, ESA, F. Paresce (INAF-IASF, Bologna, Italy), R. O'Connell (University of Virginia, Charlottesville), and the Wide Field Camera 3 Science Oversight Committee; 212: ESO; 214: ESO; 216: Alex Cherney, Terrastro.com/Science Photo Library.

All other maps and illustrations by Pikaia Imaging
www.pikaia-imaging.co.uk

Quercus Editions Ltd
Carmelite House
50 Victoria Embankment
London
EC4Y 0DZ

First published in 2015

Text by Giles Sparrow
Map illustration by Pikaia Imaging
Design layout by Stonecastle Graphics Ltd

A catalog record of this book is available from the British Library

UK and associated territories: ISBN 978-1-84866-913-0

Printed and bound in Italy by L.E.G.O. S.p.A.

10 9 8 7